土人景观设计著作系列

# 人民广场——都江堰广场案例

## People's Square: The Dujiangyan Square Case

俞孔坚 石颖 Mary Pudua 等著

中国建筑工业出版社

**图书在版编目(CIP)数据**

人民广场——都江堰广场案例／俞孔坚等著.—北京：中国建筑工业出版社，2004
(土人景观设计著作系列)
ISBN 7-112-06721-9

I.人... II.俞... III.广场－建筑设计－案例－都江堰市
IV.TU984.271.3

中国版本图书馆 CIP 数据核字（2004）第 061708 号

责任编辑　郑淮兵　张惠珍
责任设计　刘向阳
责任校对　赵明霞

土人景观设计著作系列
**人民广场——都江堰广场案例**
People's Square: The Dujiangyan Square Case
俞孔坚　石 颖　Mary Pudua　等著
*
中国建筑工业出版社出版、发行(北京西郊百万庄)
新 华 书 店 经 销
北京嘉泰利德公司制版
北京方嘉彩色印刷有限责任公司印刷
*
开本：787 × 1092毫米　1/12　印张：11⅔　字数：300千字
2004年8月第一版　2004年8月第一次印刷
印数：1—2,000册　定价：115.00元
ISBN 7-112-06721-9
TU · 5869(12675)
**版权所有　翻印必究**
如有印装质量问题，可寄本社退换
(邮政编码 100037)
本社网址：http://www.china-abp.com.cn
网上书店：http://www.china-building.com.cn

# 序

今天早上（2004.5.6）在电视新闻中，看到温家宝总理访问比利时，在接见当地的华侨和留学生时说的一番话，令我感动得泪流满面，大意如此：我希望每个中国人都能生活得好，每一个农村的孩子都能上学，每一个得病的人都能得到很好的医治。它之所以让我感动，是因为其朴实无华，没有豪言壮语，没有辉煌的前景描述，但对当代中国来说却是伟大和充满意味而值得纪念的，它的纪念性远远高于其他关于“人民”的昂扬的语言。

本书并无意将城市和景观的设计与当代领导人拉上关系，但景观确实又是政治的和社会意识形态的符号，更何况我们是在谈“人民的广场”。于是我又想到与本书主题相关的都江堰之于长城的意义，有许多学者有过精辟的讨论。我欣赏余秋雨的说法：“我以为，中国历史上最激动人心的工程不是长城，而是都江堰……它的水流不像万里长城那样突兀在外，而是细细浸润、节节延伸，延伸的距离并不比长城短。长城的文明是一种僵硬的雕塑，它的文明是一种灵动的生活。”

作为一种政治的和社会的景观，现代城市空间是要像长城那样傲然而获得其纪念性令万民供奉顶礼膜拜，还是像都江堰那样平实亲切，成为市民生活中最灵动的部分？本人的立场是不言而喻的：现代城市空间不是为神设计的，也不是为君主设计的，更不是为市长们设计的，而是为生活在城市中普通的人们设计的，这些普通的人是具体、富有人性的个体，而不是抽象的集体名词“人民”。本书通过一个案例，强调人在场所中的体验，强调普通人在普通环境中的活动，强调场所的物理特征、人的活动以及含义的三位一体的整体性。场所或景观不是让人参观的，而是供人使用、让人成为其中的一部分。呼吁场所的诗意的回归、人性的回归和故事的回归。

如何来获得场所性与人性的回归，如何来唤起人性与公民性是贯穿整个设计过程的一个思考线路。

我曾经在阳春三月，穿行于川西烂漫的油菜花丛中，偶然间误入竹林匝护的农家院落，车门的书法对联，透出无限春光和憧憬，白石灰刷就的建筑山墙露出木头的穿斗结构；窗棂和木格栅是饱经风霜的棕褐色。青色、白色和红色的大卵石界定出屋沿垂直投影的内和外，院子的一边是一口井，井也是卵石砌就的。院中一群老人正在点燃千百支蜡烛，并环绕场院走动着，脸上的安详和宗教的虔诚唤起我儿时关于奶奶的记忆。这种久违了的感觉突然间将我置身于久远，仿佛来到只知有秦不知有汉的桃源。

我曾在傍晚时分游逛于湍急的都江堰的内江两岸，感受江风送来涛声，携带着极具个性的成都话音和拥挤、热闹而不紧不慢的信息。我也曾沿都江堰众多灌渠中的一条，大大小小的鱼嘴和导水槽将来自闽江的清流引向广袤的田野，像一个致密而系统的神经网络，滋润着整个川西平原。随她们中的任何一条支流，我穿过柳丛和桃林，走进村庄，见石埠上浣衣的妹子，尽情地将顺流而下的波纹搅碎；也见到一对老人肩负着竹篓，一个篓里是婴儿，另一个篓里是瓜菜，一前一后，慢慢踏过横架在水渠上的红砂岩石板桥；村头的楠木林下是三五成群的男女，围着麻将桌，陶醉于这无异男女老少的游戏之中；那林荫投入水中，林荫下是两个儿童，用竹编的筛子，权当渔具，捕捞石滩中的小鱼。体验如此完善的灌溉系统和由此串联起来的生活，使我关于都江堰的认识已不再是一堰一坝的概念。

正是循着都江堰的水脉，我读到了大地的肌理和过程，读到了历史和故事，读到了当地人的生活，也读到了人性，而所有这些，便是我们的都江堰广场设计的灵感源泉。

而我也庆幸侯雄飞先生——当时的都江堰市委书记，一位中国人民大学的硕士研究生，一位难得的城市决策者，也读懂了我和我们的方案。记得，1999年初都江堰市人民政府开始在全国进行广场设计方案竞赛，有10家国家甲级规划设计单位参与竞赛，最终北京土人景观规划设计研究所和北京大学景观规划设计中心的方案获得第一名并中标。侯书记另有重任之后，随后的书记和市长张宁生、将显伦、张济环、万钧等都不遗余力地推动广场的实施和力求完满的建设。在长达四年的设计和实施过程中，都江堰市人民政府和规划局、建设局等部门的众多领导都给予了大力的支持，将他们的名字一一列出是困难的，但决不应该忘记余伏麟、邱崇伦、王聪、袁明的直接而重要的贡献。

本项目由土人景观和北京大学景观规划设计中心主持设计，参与工作的人很多，包括：石颖、张东、李健宏、胡海波、董涛、唐慧玲、胡东风、罗华、冯垣、魏鹏程等。除了土人的设计师们以外，艺术家郭选昌、景观工程专家姚振楠在广场实施建设中的积极贡献也是不能被忘记的。施工单位上海园林工程公司和北京土人景观生态工程公司在建设过程中克服种种技术和非技术的困难，最终完成了一项艰难的工程，给广场增加了另一层含义。

俞孔坚　都江堰广场首席设计师

北京大学景观设计学研究院教授　院长

北京土人景观规划设计研究所所长

2004年5月6日于北京上地

# 目　录

# Content Table

阅读大地
To read the land and the region

# 1　设计源于解读地域、历史和生活①

都江堰广场位于四川省成都，都江堰市（原灌县）。城市因有二千多年历史的大型水利工程都江堰而得名。该堰是我国现存的最古老而且依旧在灌溉田畴的世界级文化遗产。广场所在地位于城市中心，柏条河、走马河、江安河三条灌渠穿流城区。同时，城市主干道横穿东西，场地被分为三块，占地11$hm^2$。该地段原为大量危旧平房，1998年市政府在旧城改造时，拆出广场用地，旨在亮出灌渠和鱼嘴。1999年初开始在全国进行广场设计方案竞赛，有10家国家甲级规划设计单位参与竞赛，最终北京土人景观规划设计研究所和北京大学景观规划设计中心的方案获得第一名并中标。此后经过长达四年的设计和施工，终于在2003年5月1日基本完工。

广场是人与人交流的公共场所。一个完整意义上的场所由空间的物理特征、人的活动和历史文化涵义所构成。本案例从地域的自然和文化过程，历史，场所的现状问题和当地人的生活及休闲方式诸方面入手，理解、分析问题和解决问题，并将景观的艺术设计理解为解读地域和场所精神的过程。

## 1.1　设计目标

用现代景观设计语言，体现古老、悠远且独具特色的水文化，以及围绕水的治理和利用而产生的石文化、建筑（包括桥）文化和种植文化。使之成为一个既现代又充满文化内涵的、高品位、高水平的城市中心广场。具体功能有以下三方面：

(1)文化功能：作为都江堰市的文化体验空间

它将作为标志性的城市文化景观，集数千年文化于一体，充分体现都江堰市的地方文化和地方精神。作为都江堰工程的一部分，将内江二分为四，使千万亩良田受其滋润。

(2)休闲功能：作为市民的身心再生空间

如果将水、石和植物相交融，广场可成为一处绝佳的生态与休闲环境，可成为市民休憩、交往、公共活动和亲近自然、享受大自然的理想场所。

(3)旅游功能：作为重要的旅游节点

都江堰以其悠远的历史、功垂万世的水利工程吸引了大批的游客，而广场

---

①本文的主要内容发表在：建筑学报，2003(9):46-49. 作者俞孔坚，石颖，郭选昌

是整体中的局部，它与其他旅游点一起，全面展现了都江堰工程的气势与风采，同时还可让游人领略蜀地的气息，体验当地的市井文化。

## 1.2　设计构思来源之一：地域自然与人文景观研究与历史阅读

### 文化挖掘——水文化之精神

因水设堰，因堰兴城，水文化是都江堰市特色的来源与根本特征。为此，都江堰广场之场地特征集中体现在以下两点：

### 1.2.1　地域景观：天府之源——自然与文化景观格局

都江堰市枕居都江古堰，坐落在群峰脚下，镶嵌于蒲阳、走马、江安、柏条四河之间。西北古堰雄姿，群山环峙；东南平畴万里，天府良田无垠。其水由西北向东南而行，畅游百川，泽万顷良田，奔腾呼啸，气势磅礴。放射状的水网奠定了天府之国扇形自然景观格局的基础。都江堰是天府扇面的起点，而广场则为都江堰市的扇面核心或称“水口”。它泽被天府，沃野万里，是政治、经济和文化发展之依赖，这里是天府之国水文化之发源地，是谓“天府之源”。

### 1.2.2　阅读历史：饮水思源——以治水、用水为核心的历史文脉及含义

（1）治水的渊源

都江堰水文化的始祖首推大禹。大禹治水，以岷江为首功。《尚书·禹贡》明确记载“岷山导江，东别为沱”。都江堰一系列大大小小的分水鱼嘴可视为一系列的“东别为沱”。大禹治水顺应自然，乘势利导，禹之功“披九州，通九泽，决九江，定九州”，虽大多分布于黄河流域，但也是李冰治水思想的体现，也是都江堰得以产生的远古历史背景。

都江堰的历史可远溯至神话中的“鳖灵治水”。《蜀王本纪》载：望帝时“玉山出水，若尧之洪水，望帝不能治，使鳖灵绝玉山，民得安处。”

都江堰真正为多数人公认的始创者是秦昭王时的蜀郡守——李冰。他“乘势利导，因时制宜”，“凿离堆，避沫水之害，穿二江成都之中，此渠皆可行舟，有余则用溉浸，百姓享其利”。李冰当之无愧是世界上最伟大、最杰出的水利科学家，但都江堰却是一个千秋功业，它凝聚了数千年蜀人的辛劳。继李冰之后，各朝代的人都对都江堰进行了修复与改造，技术上不断得以更新。

（2）种植文化

昔古蜀之地，“江水初荡橘，蜀人几为鱼”。都江堰建立后“又灌溉三郡，开稻田，于是蜀沃野千里，号称‘路海’，……水旱从人，不知饥馑，时无荒年，天下谓之‘天府’”。随都江堰工程的进一步完善，成都平原处处皆为人间乐土，沟渠纵横，阡陌交错，地无旷土，已到了“天孙纵有闲针线，难绣四川

解读历史
To explore the history

百里图”的佳境。汉化后的种植文化将种植与养殖统一，稻鱼结合，自成体系。

种植文化使蜀“人杰地灵”，成为政治经济的重地，也为后世种植、渔业等的发展打下了基础，并激发了我们的设计思路与灵感。

（3）植根古蜀的建筑技术

古蜀文化，重视阴阳五行说，主张人与自然的和谐统一。都江堰建设中的基本特征也是如此，“深掏滩，低作堰”，“遇湾截角，逢正抽心”等，主要工程都顺水势而行，乘势利导，因时制宜。

都江堰工程中的若干重要技术，如笼石技术（竹笼卵石及后来的羊圈－木桩石笼工程）、鱼嘴技术、火烧崖、石凿崖技术、都江堰渠首和有关河渠上的若干索桥的建筑技术，都具浓重的地方水利风格，富有民族文化特征。

另外川西民居及宗教建筑也有鲜明的特点，特别是对竹木石材的应用及艺术处理都有其浓厚的地方特色。

（4）石文化

古代蜀人有崇拜大石、崖石的原始宗教意识。石犀、石马是具有地方特征的神物，蜀人认为犀牛神可战胜水神。“李冰于玉女房下白沙邮作三石人，立三水中，与江神约：水竭不至足，盛不没肩”及“作石犀五枚”，“以厌（压）水精”，皆适应了蜀人的大石崇拜意识，且有阴阳五行说的内涵，从五行相克的关系看，石属土，土胜水，石神有镇水的含义。

（5）水的其他衍生文化

都江堰因水而生的文化很多，如神话、祭祀文化、景观及历代文人墨客留下的诗词歌赋等，这里只提一下都江堰的开水节，它源于远古对水神的祭祀，后改祭李冰。每年“都江堰水沃千川，人到开时拥岸边；喜看杩槎频拆处，欢声雷动说耕田”。古老的庆典民俗相沿千年以上，现已成为都江堰市极具特色的传统节日。它增进了人们对水文化的认识，并永记李冰之功。

### 1.2.3 市民休闲行为之特点

时代的发展使得许多城市生活节奏愈来愈快，而这里却处处有别地少见的悠闲景象。成都有“最悠闲城市”之美誉，而其中尤以都江堰为最。这里的人们喜欢夜啤酒、饮茶、打牌、聊天、遛鸟、练功，喜欢悠哉游哉的日子，颇有“长夏无事，夕阳西下，明月东升。搬个小板凳，沏上壶不浓不淡的茶，聚几个不衫不履的人，说几句无拘无束的话”的味道，只不过此时此地的人们不分春夏秋冬，不辨白天黑夜。

### 1.2.4 作为旅游地之潜力

（1）广场是都江堰整体之一个部分

望外江、登玉垒山、观离堆……，而不到广场则无法体味都江堰工程的全

感知文化和地方之神
To perceive the vernacular landscape and *genius loci*

部意味。另外，都江堰因水而兴，余秋雨曰“看云看雾看日出各有胜地，要看水，万不可忘了都江堰”。而广场又是水文化的浓缩，不可不到。

（2）广场是体验市井文化的场所

形形色色的市民在此消闲、聚会，是都江堰市民生活方式的集中展现，游客品异地风情的好奇心可大大得以满足。

（3）广场是城市的核心所在

其周围商业街、商业大楼及文化活动中心等遍布，而其自身亦环境幽雅，独具特色，势必会吸引大量游人。

## 1.3 设计构思来源之二：场地问题的分析

景观设计的要旨是解决问题，都江堰广场现状有许多问题必须通过设计来解决：

(1)城市主干道横穿广场，将广场一分为二，人车混杂。

(2)用于分水的三个鱼嘴没有充分显现，本应是最精彩的景观，却被脏乱所埋没。

(3)渠道水深流急，难以亲近，其中一段已被覆盖。

(4)广场被水渠分割得四分五裂，不利于形成整体空间。

(5)局部人满为患，而大部分地带却无人光顾。

(6)多处水利设施造型简陋，破败不堪。

(7)大部分地区为水泥铺地，缺乏景观特色与生机。

(8)周围建筑既无时代气息，也无地方特色。

## 1.4 设计构思：解决问题与营造场所

基于以上地域景观和历史渊源的分析，以及场所问题和特征的研究，提出以下设计构思：

天府之源，编鱼嘴竹笼；稻香荷肥，织沟渎阡陌。

饮水思源，循上下千载；秉鳖灵志，续李冰功业。

投玉入波，愿与神为约；抽心截角，倡因势利导。

凿穿通渠，使人车分流；水帘茶肆，尽都市闲情。

水车提波，使触手可及；鱼嘴喷碧，皆水神石态。

青石导流，得整合场地；暗渠复现，还地方灵魂。

### 1.4.1 广场主体构思

天府之源，投玉入波；鱼嘴竹笼，编织稻香荷肥。

体验生活

To experience the living of the local people

解剖场地
Understanding the site

在广场之中心地段，设一涡旋形水景，意为“天府之源”。中立石雕编框，内填白色卵石，取古代“投玉入波”以镇水神之意，又为竹笼搏波之形，同时喻古蜀之大石崇拜，金生水，土（石）克水，相生相克，体现治水之要旨。石柱上水花飞溅，其下浪泉翻滚，夜晚彩灯之下，浮光掠金。彩灯光束呈杩槎之形，尤为动人。

水波顺扇形水道盘旋而下，扇面上折石凸起，似鱼嘴般将水一分为二、二分为四、四分为八……细薄水波纹编织成一个流动的网，波光粼粼，意味深远，令人深思；蜿蜒细水顺扇面而下，直达太平步行街，取“遇湾截角，逢正抽心”之意。

广场的铺装和草地之上是三个没有编织完的、平展开来的“竹笼”。竹篾（草带、水带或石带）之中心线分别指向“天府之源”。中部“竹笼”为草带方格，罩于平静的水体之上，中心为圆台形白色卵石堆。东部“竹笼”则以稻秧（后改为花岗石）构成方格，罩于白色卵石之上，中置梯形草堆（后改为卵石堆）。西边“竹笼”则是红砂岩方格罩于草地之上。

这些没有编织完的竹笼之平展方格同时象形于水利灌溉之下的种植文化（早在汉代石刻上就有种、养殖之地块分割图）。

### 1.4.2 问题的解决对策

（1）整合场地

针对水渠将广场分割的现状，以向心轴线整合场地。轴线以青石导流，喻灌渠之意，隐杩槎之形。可观、可憩、可滋灌周边草树稻荷。同时在各条水渠之上将水喷射于对岸，夜光中如虹桥渡波。

（2）人车分流

干道处为避免人车混杂，以下沉广场和地道疏导人流。广场北侧半圆形水幕垂帘，茶肆隐于其中（后取消）；南端水流盘旋而下，以扇形水势融于地面并成条石水埠之景。

（3）强化鱼嘴

四射的喷泉展现了分水时的气势，突出了鱼嘴处水流的喧哗。水落而成的水幕又使鱼嘴及周围景致若隐若现，独具情趣。灯光之下，如彩虹飘带，挂于灌渠之上。

（4）分散人流

广场四处皆提供小憩、游玩之地，市民的活动范围将不会再局限于现有的小游园处。

（5）增强亲水性

设计后整个广场处处有水，注重亲水性的处理，重点有以下几方面：

内江处水车提水，引水流于地面，游人触手可及。

广场南部以展开的竹笼之形，阡陌纵横之态，引水以入，市民尽可在其间游玩。

蒲阳河上暗渠复现，但水薄流缓，人可涉而过之，倒影入水，人水交融（后改为旱地喷泉）。

（6）重塑水闸

利用当地的石材——红砂岩，将闸房建筑进行改造。罩以红砂岩框，上悬垂藤植物，周围以白卵石铺装，兼悬水帘，将水闸以一种独具特色的建筑融入广场的环境与氛围。

（7）创建生态环境

广场上水流穿插、稻香荷肥、绿草如茵、树影婆娑，一改以往水泥铺地的呆板，营造出一片绿意与生机，成为都江堰市一处难得的生态绿地，市民身心再生的极佳空间。

（8）营造生活情趣

广场的设计因袭当地的市民文化和村落街坊共赏院落格局，注重意境的创造，强调精致的细节。茶肆遍布，处处隐于林中；南端小桥流水，别具情趣；阡陌中或石或水，妙趣横生；疏林草地上，座椅遍布，市民或坐或卧、或读或聊；青石渠，红砂路，水、树、人融于一体。

（9）交通体系

将城市交通干线移出广场区域，限制穿越广场的车流。未来的停车场最好位于拟建博物馆一带，这样既便于参观博物馆和通达广场，又可减少车流对广场的干扰。

（10）周边建筑

要注重风格的统一，并强化地方特色和时代感。拆除杂乱建筑，有重点、有目的地进行建设，同时要强化建筑周围环境绿化的效果。以凤凰宾馆为例，采用具地方特色的红砂岩框将其进行改造处理，古中有新，又很有时代感。

（11）河畔处理

广场临水段预留不少于8m的步行道和草地，用作防洪抢险通道。同时建议加强沿河两侧的整体绿化工程，并延伸至下游，重建扇形绿色通道，以充分发挥水的生态作用，将其创建成都江堰市集休闲、娱乐、生态功能为一体的绿色生态走廊。

（12）灯光及广告

灯光是为夜晚增添情趣和闪光点的关键。都江堰市气候较好，夜间活动可持续到很晚，因此可将广场设计为不夜之地，除一般照明外，要以艺术照明的手段点缀其间。

广场未来作为都江堰的核心和游客的集中地，在恰当的部位设置广告牌，有助于让游人了解都江堰的发展及工商业状况。部分广告牌可和灯柱结合，多媒体广告牌可设在现电视大楼之东侧的裙房屋顶。

解剖场地
Understanding the site

### 1.4.3 广场的艺术设计

广场的艺术设计来源于对地域自然和历史及文化的体验和理解，也来源于对当地生活的体验，综合起来是对地方精神的感悟。李冰治水的悠远故事，竹笼和杩槎的治水技术，红砂岩的导水渠和分水鱼嘴的巧妙，川西建筑的穿斗结构和红木花窗，阳春三月走进川西油菜花地中的那种纯黄色彩和激动心情，还有那井院中的卵石和竹编的篱笆，老乡的竹编背篓，打牌或静坐的老人和青年，围坐在麻将桌边的姑娘，麻辣酸味的鱼腥草……一切都在为这场地的设计提供语言和词汇。除了上述“投玉入波”主体景观外，以下是其他几个重点艺术处理的解释：

(1)导水漏墙：源于竹笼和导水槽的艺术集中体现在广场中部斜穿广场的石质格栅景墙上。该景墙采用10cm × 10cm镂空，斜向方格肌理，顶部为导水槽。该景墙既起到分割广场空间的作用，同时，由于其为通透的漏墙，使广场分而不隔，丰富了广场的空间和景致。近百米长的格栅景墙强化了南北向的轴线关系，这条轴向北源于闽山豁口，即宝瓶口，往南延伸入未来太平街步行街，中间则是广场的主体雕塑“投玉入波”。景墙与“投玉入波”之间通过3根高3m的灯柱来虚接。景墙与太平街的联系通过一个井院式卵石广场作转折。

(2) 杩槎天幔：广场设计之初，时逢阳春三月，天府之国，见菜花遍地；或行游其中，激情荡漾，流连忘返。再寻觅三千年前文脉，得杩槎治水之妙道。因此遍插铜柱，斜立有致，侧观如杩槎群，上悬黄色天幔，犹如遍地黄花。夜光之下，更为灿烂。

(3) 灯柱及栏杆系列：采用相同的格栅语言，在广场上布置了一系列灯柱，构成广场的主要竖向结构。外为花岗石材质，内衬毛玻璃，夜晚可形成独特风景。竹笼的编制格式语言体现在广场设计的任何一个细部，包括所有临水栏杆和过江的两座吊桥。

(4) 微空间设计：生活和休闲本身是艺术，都江堰人的休闲方式为公共场所的空间设计提供了不尽的艺术灵感。在广场的北端，历史上是行舟拴缆绳的地方，残存的水泥墩自然成为休闲老人聚首耍牌的场所，这是当地人休闲方式的表达。为再现和再生这样的场所，满足当地人的休闲需要，设计了约5m见方的近人尺度的空间，并为这种休闲提供了充足的座凳和观赏游玩的场所。广场西南角是一处下沉式演艺广场，观演场所打破了一般的做法，而是通过叠石和乱石、环境艺术品及种植，设置了多样化的空间；广场南端利用地势，形成下沉跌水空间，并和荫棚长廊相结合，构成一条林下休闲场所。导水漏墙的落水处为一卵石铺垫的院落，柱廊围合，中为水井，吸纳漏墙跌水。自院内循墙北望，但见白水一注自闽山而来，“天府之源”意在其中。

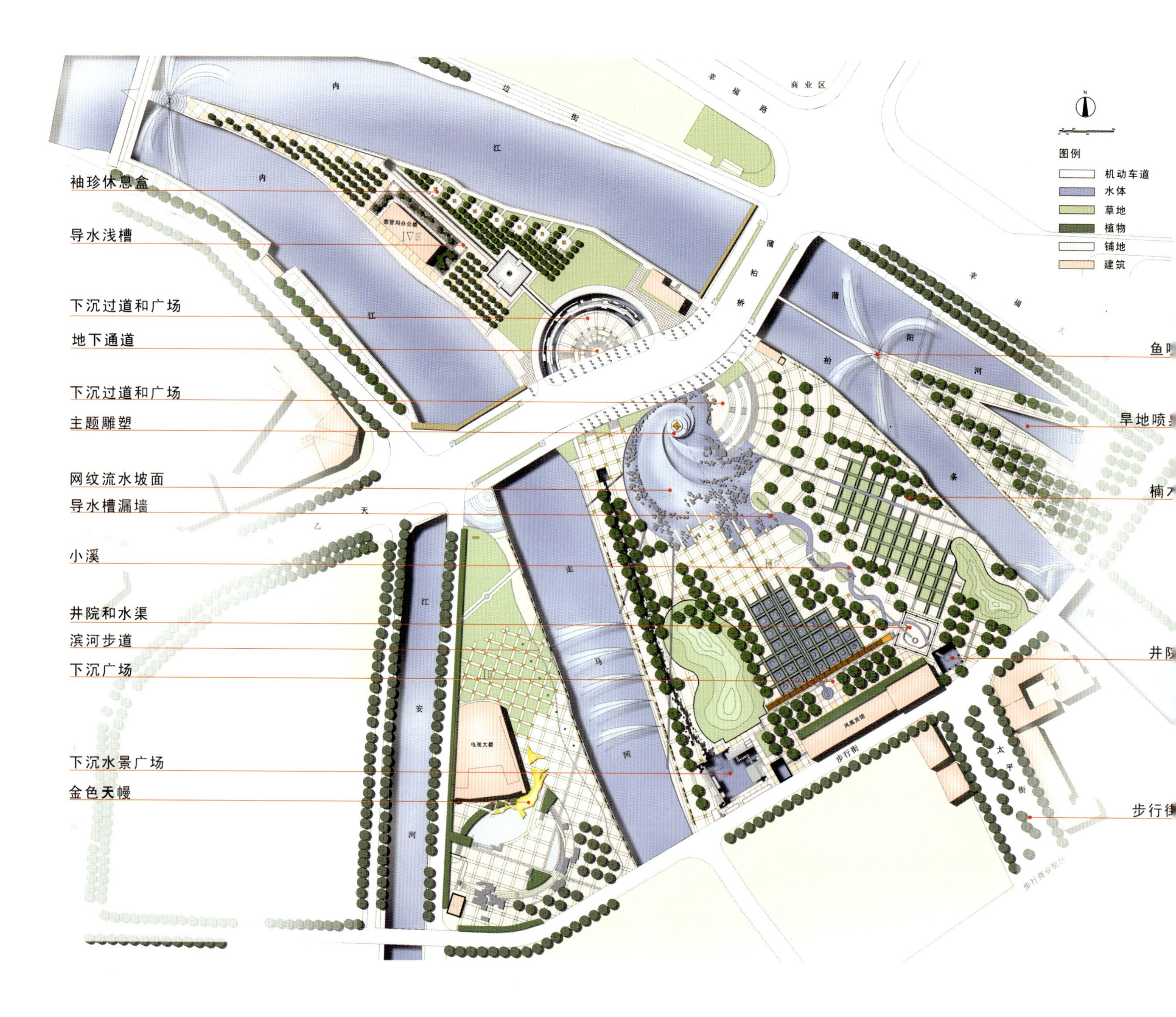


都江堰广场总平面图:一个展开的竹笼，扇形。视觉焦点是一个源于竹笼原形的雕塑。中部为一条斜向轴线，该轴向北指向闽山豁口中的都江堰，往南连接未来的步行街

The master plan of the Dujiangyan Square: an unfolding bamboo basket, radiating from the focal point where stands the symbolic sculpture on a whirl shaped water feature. A diagonal axis made of carved stone wall and an meandering creek visually dominants the square that connects the pedestrian street down at the south, and the Dujiangyan Weir, the symbolic water source of the Chengdu Basin up in the north kilometers away

都江堰广场中心区（II 区模型）
The model of the central part of Dujiangyan Square

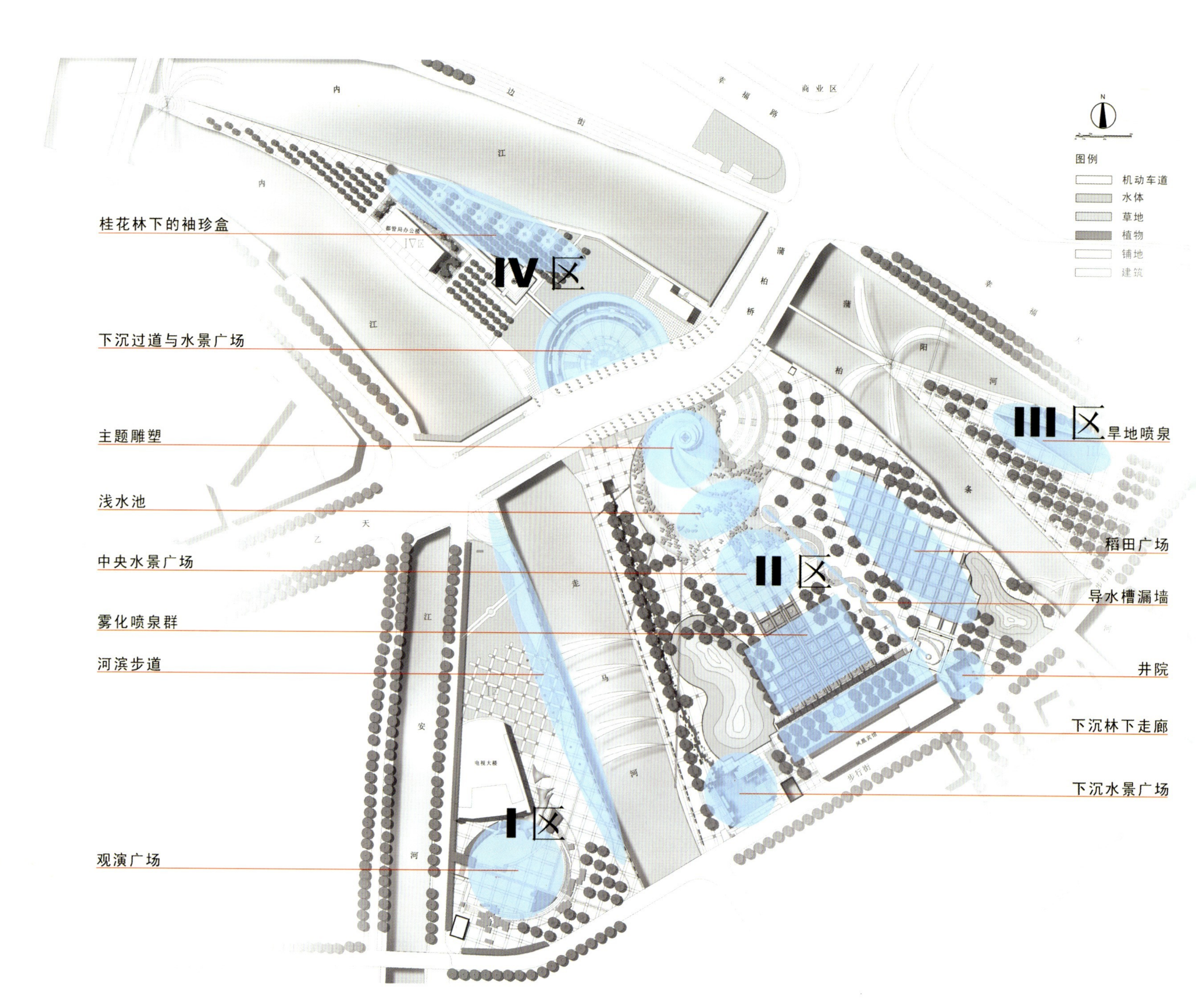


分区与多样化的公共场所
The phasing and diversity of the square

# 2 寻回广场的人性与公民性[①]

城市广场本来是人性与公民性的体现，是人与人的交流场所，是每个人参与社会获得认同并以之为归属的场所，本质上是一种政治景观。然而，城市广场的人性与公民性早在欧洲巴洛克时代就已丧失，而在当今中国“城市化妆运动”中更是如此，一元化的社会意识使广场的人性与公民性只能在一些乡土景观中尚有遗存。随着中国日益走向民主政治和平民化时代，城市设计，特别是广场设计应首先召回其人性与公民性，用进步的物质环境，积极推动社会意识的进步。经历三年实施完成的四川都江堰广场探索了如何通过设计多元化的空间，可参与交流和聚会的场所，增强场所的人性化、亲水性，并增强场所的认同感与归属感，来实现广场的人性与公民性的回归。

## 2.1 广场的原义：唤起人性与公民性的场所

Paul Zucker（1959）认为：广场是使社区成为社区的场所，而不仅仅是众多单个人的聚集，……是人们聚会的场所，是人通过相互接触和交流而被教化并被赋予人性的场所，广场给人们提供了一个躲避交通事故危险的庇护所和安全岛，是在繁忙的城市交通网络中，使人可以重获安全和自由的一个场所。

人文景观学者Jackson指出，广场决不应仅仅理解为一个环境和展示性的舞台，广场的内涵是极其丰富的，它曾经是，并仍然是当地社会秩序的显示，是人与人，市民与当权者之间关系的反映。广场使个体在社区中的地位和作用得以显现，它使每一个人在社会中的政治、信仰和消费归属和认同得以彰显，并使这种归属和认同得以强化。公共广场不仅仅是一个供人休闲和唤起人们环境意识的场所，它也是唤起公民意识的场所。

广场本质上是一种政治景观，是社区中的个体参与公共活动、参与社会，并在其中显示其角色的场所。在西方，广场的这种属性的起源最早可以追溯至古希腊军队的例行立队集会，士兵们围成圈来讨论共同关心的问题。每个士兵轮流步入圈内，自由表达自己的建议，讲完后再退回到队伍中，然后另一个士兵再踏入圈内发言，这个“圈”就叫Agora，意即集合(assembly)。随着时间的推移，这种“圈”被当作合格的公民的聚会，在此对共同关心的问题自由辩论，各抒己见。这种“圈”被认为是西方公共广场（public square）的词义的源头（Jackson，1984，

① 本文主要内容发表在：新建筑，2004。作者俞孔坚，万钧，石颖

寻找人性与公民性的场所：乡土景观中的场所
People's places in vernacular landscapes

人性与公民性的丧失：
神权与君权下的一元化场所（圣彼得广场、天坛和某广场上的人民）
Places of oneness:loss of humanity

P.18）。但广场的存在却可以至少追溯到有人类居住史迹的那一刻。

早在原始母系氏族公社时代，人类在建筑属于自己的私密空间或庇护所的同时，围合了一个用于讨论公共事务的场所。如在距今6000年的西安半坡仰韶文化遗址中，在3000多平方米的场地中分布着46座房屋，总体上呈不规则圆形，分为两片，围合着中间的“广场”和大房子；更典型的例子是陕西临潼姜寨遗址，在18000m$^2$的居住范围内，以中心广场为中心，分布着五个组团的房子，每个组团的中心都有各自的中心和一座大房子，四周由20多座房子围绕着，而大小房子的门都开向中心广场。中间这个广场代表了“公家”，是集体的体现，在这个场所里，人人是平等的，没有人比其他人优越，也没有人比别人低劣。在这里诞生了社会，一个由可以互换角色的、独立的个体所构成的社会。可以说，允许进入这个公共的场所，就意味着有共同参与政治、讨论公共事务的权利。这种社会性、民主性、平等性和可以互换角色的对称性是公共广场的原始含义（Jackson 引自 Jean-Pierre Vernant，1984，P.19）。

如果说广场使社区成为社区，使社会成为社会，那么，广场实际上也使人成其为人。广场本身是人作为群居的社会性动物的本质属性的反应，正如人需要私密的庇护空间一样，人需要作为交流空间的广场。对广场和交流场所的需求我们甚至可以在人类的远祖同类猕猴和猩猩那里得到证实。

广场的人性与公民性的特征在古罗马共和国的广场中，在中国古代和近代的村镇广场和广大的庙会场所中，以及在村口的大树下和镇中心的水井边等公共场所，都得到延续和张扬。正是在这些公共的场所，当地的人民找到了认同和归属，认识了自己存在的价值，获得在社区和社会中的地位。如果说认同和定位使场所具有意义，才使生活其中的人生活具有意义的话，这样的能体现人性与公民性的广场，是社区与城市不可或缺的。

## 2.2　广场的失落

对广场的讨论近一两年来成为城市设计讨论的一个热点，主要是对近年来的广场作为城市化妆功能的讨论和反思（如俞孔坚，吉庆萍，2000；俞孔坚，李迪华，2003；郭恩章，2002；金经元，2002；王建国，高源，2002；段进，2002；牛慧恩，2002）。在作者看来，目前广场建设热潮中的一个核心问题是广场的人性与公民性的丧失，因此，现代中国广场设计的首要任务是重归人性，重归广场的公民性，也就是要召回广场的原本含义。

当然，广场的人性与公民性的丧失并非从今天才开始，也决不是从中国开始。从15世纪中叶开始，罗马教皇尼古拉五世(Nikolaus V，1447-1455年在位)就开始了将一个平民的城市改造成神圣的城市的计划，这一改造计划一直延续到16世纪末，历经100多年的时间（贝纳沃罗，2000）。在这一广场的神圣化和纪念化过程中，艺术家成为教皇们实现宏伟目标的工具，一向被教科书顶礼膜拜的艺术家们如米开朗琪罗、拉斐尔被邀请设计广场和广场上的建筑，艺术的完美和宏伟壮丽成为第一、甚至惟一标准，其中最典型的代表是圣彼得广场。超人尺度的巨形围廊、为展示圣彼得教堂的极度宏伟而设的宽广的空间和轴线，使每一个来到广场的人都被压迫和缩小到最卑微的程度，广场上的喷泉和雕塑仅作为纪念碑和观赏的对象，没有树荫，没有座椅。在这里，人性和公民性被彻底扼杀。

17世纪下半叶，巴洛克这一有效地谋杀人性和公民性的手段同样被法国国王路易十四用于凡尔赛宫及巴黎城市的设计，后来又在1791年经法国建筑师L' Enfant传到了美国，并继而在19世纪末和20世纪初在美国的“城市美化运动”中出尽了风头（俞孔坚，吉庆萍，2000；俞孔坚，李迪华，2003）。Jackson (1984)因此感叹由于艺术主导的设计的流行，美国广场中曾经存在于乡土民间小镇公共场地上及乡村教堂广场上的公民性和人性因此丧失。好在美国一向以对人性和公民性的尊重作为城市建设的价值取向，因而“城市美化运动”在美国本土并没有持续多久，很快被广大专业人士和民众所唾弃。然而，几乎就在同时，违背人性与公民性的城市广场和公共空间设计却在欧美帝国主义的殖民地大行其道，在20世纪30年代又被欧洲新一代的独裁者所借用。

很显然，人性与公民性是一个健康的社会和政体在城市公共空间中的反映，反之亦然。作为一个走向世界，开放、民主的中国新社会，在“三个代表”精神指引下的新一代政府部门理所当然应将人性与公民性还给城市公共场所。

## 2.3 重归广场的人性与公民性：都江堰广场案例

作为一个实践案例，都江堰广场在唤起广场的人性与公民性方面，着重体现在以下几个方面：

### 2.3.1 多元化的空间

无论是欧洲巴洛克的广场和城市设计，还是中国封建都城（如紫禁城）及神坛（如天坛），都是通过强烈的轴线和占绝对统治的一个中心（如雕塑，喷泉）来形成一元化的空间，以表达一种绝对权力的存在。这往往使其成为与没有设计师设计的乡土广场，或日常生活的社区广场的一个主要区别。还广场以人性与公民性就必须首先打破在封建和极权意志下形成的一元化空间形式。都江堰广场也有一个作为中心的主雕塑，高30m，起到挈领被河水和城市干道分割的四个板块的作用，与雕塑成一直线的是一道导水镂墙，构成一条轴线。但这一中心和轴线更多的是起到空间组织联系和视觉参照的作用，并没有损害广场空间的多元化，形象地说，主题雕塑和镂墙在这里是个“协调者”而非“统

都江堰广场：营造多元化的空间
Spaces of diversity

都江堰广场：营造参与交流和聚会的场所
A place for meeting and participatiation

治者。”

利用场地被河流和城市主干道切割后形成的四个区块，形成四个功能相对有别，但又互为融合交叉的区域，动中有静，静处有动，大小空间相套，既有联系又有区分：

I区，以观演广场为主，设有舞台，常为热闹演艺场所和小群人晨练场所；同时又有滨河休闲带供使用者静处、散步或欣赏河流波涛，并有林下休闲区，供来自南部居民区的使用者小聚聊天，遛鸟晨练。

II区，以水景和平地广场为主，早晨和傍晚常为多数人的群众性体育活动和舞蹈最爱处。平时则是儿童戏水的乐园，这里有雾泉、高塔落水、坡面流水、卵石水池。这里也是观光客的最佳去处，留影者可以找到许多奇妙的景观作背景。在南部和西南部设安静的林下休闲广场和下沉式水池空间，一条蔽荫长廊将其与热闹的北部分开，大量的树荫、座凳、安静的空间，最受邻里居民的青睐。西侧临河，与对岸I区的滨河休闲走廊相呼应，设大量石条凳，以供休憩、观赏河水。

III区，以一组可参与的旱地喷泉为主，吸引大量儿童和大人观赏和游玩。南、西两侧为樟木林，为游客提供大量的林下休息空间。西侧滨河带则同样提供石条座凳，近观河水。

IV区，桂花林下的袖珍空间，5m × 5m见方，最宜三五成群的麻将客和耍牌者，而这正是当地民众的喜好。这种空间设计源于原场地中土坑的尺度，而当时现场考察发现，这些土坑，加上曾用于拴缆绳的水泥墩，是牌友们最喜爱的场所。

在II区和IV区之间，则是一下沉广场，与隧道相结合，沟通两区，叠瀑环绕，形成另一种体验空间。

### 2.3.2 参与交流和聚会的场所

广场的设计从总体到局部都考虑到人的使用需要，考虑到作为人与人交流和聚会公共场所的需要：

（1）观演式交流：在I区观演舞台的设计中，演出者、观众、伴奏和后台排练，都通过景观设计的空间处理手法，形成既有联系，又有分隔的空间。舞台上是“金色天幔”雕塑，它既是舞台布景，又是独立的雕塑，同时是演出时伴奏队的遮护。舞台背后有竹丛围合，其后是后台的排练场，一个绿色豁口将前后台联系在一起。舞台前景是一组涌泉。隔着下沉式广场，与舞台遥相呼应的是台阶和高低叠石构成的观众席。一些叠石一直延续到下沉广场之中。一切都在风景之中：有表演活动时，这是一个风景中的舞台——回到风景的原始含义；平时则是人们日常活动的背景，使最普通的体育活动如舞剑、扇子舞，甚至是斜穿广场的人，也成为一种值得观玩的景致，人的活动融入到场所之中。

（2）集体自由交流：II区则为不同时段和不同人群提供了更为灵活多样的交流与聚会机会。漏墙、水景、楠木林和草地，定义了多种富有情趣的空间。早晨是集体太极拳、舞剑和各种不知名的群众性集体活动的场所；傍晚，则可以看到在音乐的伴奏下，交谊舞爱好者的翩翩起舞和成群的围观者。这种集体自由式参与和交流还以水为媒进行，它发生在II区的雾泉、浅水池和III区的旱地喷泉中。

（3）小群体交流：II区南侧的樟树林下，IV区的桂花林“盒子”空间，最适于三五成群的牌友和聊天休闲者的驻留。

### 2.3.3 人性化的设计

都江堰广场从多个方面实现人性的设计，包括：

（1）提供荫凉：结合地面铺装和座凳，在四个区内都设计了树阵，在瞬时人流量较大的I、II、III区用分枝点较高的楠木和樟树，而在小群体交流为特征的IV区，则种以分枝点较低的桂花林。

（2）座凳与台阶：广场上在合适的地方，包括广场和草地边沿、水际、林下设置大量的条石座凳，让以休闲著称的当地人有足够的休憩机会。台阶和种植池也是最好的座凳。

（3）提供“瞭望”与“庇护”的机会：理论和实验观测都表明，人在公共场所中普遍存在“窥视”的偏好（俞孔坚，1998）。看和被看是广场上最生动的游戏，在林荫中和隐蔽处，在广场和草地边缘，是最佳的“窥视”场所，因而是设置桌椅，供休憩的合适场所。而在明处或广场中央则设计活跃的景观元素，如喷泉和水体，吸引人的参与，使其无意间成为被看的对象和“演员”。

（4）避免光滑的地面：所有铺装地面都用火烧板或凿毛石材。

（5）普适性设计：广场的设计考虑各种人的使用方便，包括年轻人、儿童、老人、残疾人。

（6）尺度转换：一个11hm²的广场尺度是超人的，如何通过空间尺度的转换使之亲人宜人，是本设计所面临的一大挑战。本项目主要从四个方面实现空间的尺度转换：

第一，通过30m高的主体雕塑，使一个水平二维广场转化为三维视觉感知和体验空间；

第二，通过斜贯中心的长达100多米、高达2～8m的导水漏墙和灯柱、廊架以及乔木树阵，进一步分割空间，形成分而不隔的流动性空间体验；

第三，通过下沉广场，形成尺度适宜的围合空间。分别在I区，II区和II区与III区之间的地下通道处，设计了四个下沉式小广场；

第四，用高达3m左右的灯柱、雕塑（如“金色天幔”）和小型乔木（如桂花林和竹子），使广场空间和人体之间的关系进一步拉近。

都江堰广场:人性化的设计
The making of human space

都江堰广场:可亲可玩的水景设计
To make water playful

## 2.3.4 可亲可玩的水景设计

玩水是人性中最根深蒂固的一种。水景的丰富多样性和可戏性是本广场设计的一个主要特色。分割广场的三条灌渠，水流湍急，波涛汹涌，只可观不可亲不可玩。见如此好水只穿流而过，当地人和设计者都感可惜。所以，设计之初的一个重要设想是提河渠之水入广场，使人触手可及。为此，一次性从河中提水，从30m的“竹笼”雕塑跌落，经过有微小“鱼嘴”构成的坡面，旋转流下，水流经过时编织出一张网纹水膜，滚落浅水池中。池中大小卵石半露出水，如岷江河床上的浅水滩。从水池溢出的水又进入蜿蜒于广场上的溪流，一直流到广场的最南端，潜入井院之中。坡面上、浅池中、溪流中和井院内，都有少年儿童尽情嬉戏其中。

水面离广场铺装面近在寸许，人皆撩水为趣。步道栈桥穿越其上，步者如履镜面。

II区的雾泉、III区的旱地音乐喷泉和IV区的跌瀑，都试图实现人与水的亲切交融，充分体现都江堰的水特色。

跨越II区的导水槽使本来的水利工程设施成为一种独特的景观元素，一道银色的水流似乎自天而降，跌落到南端的井院之中，成为儿童戏耍的又一天堂。

此外，横穿广场II、III区的浅水道，再次把水的亲切与缠绵带给每一个流连于广场的人。

## 2.3.5 归属感与认同

广场是为当地人的，为当地日常生活的人的。因而广场的形式语言、空间语言都从当地的历史与地域文化中，以及当地人的日常生活方式中获得。关于这方面，已在另文讨论，此不作赘述（俞孔坚等，2003）。

总之，城市景观是社会意识形态的反映（Cosgrove，1998），人们的价值观，政治权力和国家形态，人和人的关系都直白地陈述在城市景观中，特别是城市广场这样的公共场所中。中国人的天坛和玛雅人的轴线是震撼人心的，但那是为神设计的；巴洛克城市广场和轴线是辉煌的，但那是属于帝王和君主们的，是一元政治和社会形态的体现。走过五千年一统社会的中国，终于迎来了平民化的时代，但一元化的意识将仍然顽固地存在很长一段时间。如果说人可以创造环境，而同时环境造就人和人的意识的话，那么城市设计师的重要社会责任就应该通过城市公共场所的设计，促进社会向多元化和民主时代的发展，首先应该还城市广场以人性与公民性。

## 参考文献


Cosgrove, Denis, E.1998. Social Formation and Symbolic Landscape. Madison: The University of Wisconsin Press

Jackson, J. B. 1984. Discovering the Vernacular Landscape. New Haven: Yale University Press

Zucker, Paul. 1959. Town and Square from the Agora to the Village Green. New York: Columbia University Press

贝纳沃罗.L.著，薛钟灵等译. 2000. 世界城市史. 北京：科学出版社

段进. 2002. 应重视城市广场建设的定位、定性与定量，城市规划,(1):37~38

郭恩章. 2002. 对城市广场设计中几个问题的思考. 城市规划,(2):60~63

金经元. 2002. 环境建设的“政绩”和民心. 城市规划,(1):31~35

刘致平著，王其明补. 2000. 中国居住建筑简史. 第二版. 北京：中国建筑工业出版社

牛慧恩. 2002. 城市中心广场主导功能的演变给我们的启示. 城市规划,(1): 39

王建国，高源. 2002. 谈当前我国城市广场设计的几个误区. 城市规划,(1): 36~37

俞孔坚，吉庆萍. 2000. 国际城市美化运动之于中国的教训(上，下). 中国园林,(1):27～33,(2):32～35

俞孔坚. 1998. 理想景观探源:风水与理想景观的文化意义. 北京：北京商务印书馆

俞孔坚. 1999. 谨防城市建设中的“小农意识”和“暴发户意识”. 城市发展研究,(4):52～53

俞孔坚. 2002. 城市公共空间设计呼唤人性场所. 见：中国建筑学会主编. 城市环境艺术. 沈阳：辽宁科学技术出版社，4～6

俞孔坚，李迪华. 2003. 城市景观之路——与市长们交流. 北京：中国建筑工业出版社

俞孔坚，石颖，郭选昌. 2003. 设计源于解读地域、历史和生活——都江堰广场. 建筑学报,(9):46～49

都江堰广场:归属感与认同感的营造
To create a sense of place and identity

汉代石人：用来衡量水位
The figure used for measuring water level, from Han Dynasty about 2000 years ago

## 3 评论：都江堰广场，一个叙事场所①

土人景观（北京土人景观规划设计研究所）创作的“都江堰广场”讲述了一段古老的历史——作为世界文化遗产的都江古堰的治水历史。我作为一个对该项目的评论者，对它的所在地充满了好奇。它具体在哪里？它的地域背景如何？它过去的历史是什么？它现在是个什么样子？它将对未来有何影响？这一切都对我充满了好奇。

作为分析评论，《都江堰广场》这篇文章将讨论该项目的文化背景，它所表达的内涵，该项目对当代景观设计学中水元素应用的回顾和分析，以及它在场所营造和城市再生方面的意义。

### 3.1 文化／历史背景

为了更好地理解土人景观的“水文化广场”这一项目，简短回顾一下有关文化历史是很必要的。都江堰，旧称灌县，坐落在四川省会成都西北60km处。四川不仅是个文化大省，更是中华古文明的一个重要发源地。四川被称为“天府之国”，根本原因在于古代灌溉农业时期她富饶的农业资源。这种富饶的农耕历史为现代四川的地域特色提供了深厚的基础，这当然也体现在都江堰市。

从历史角度讲，更为重要的是，在秦朝时期的土地改革，使得大量的荒地被开垦成耕地，从而促进了经济的发展和供养了大量的军队。这成为中国古代能够统一的重要基础。都江堰在古代中国意为“首府河大坝”，它是公元前3世纪中叶（战国初期），在欧亚大陆东半部建造的最为著名的一个巨大的、经过仔细规划的公共工程[1]。

它被建造在扬子江的支流岷江之上。这项工程的建造者——当地的军政首脑（蜀郡守）李冰把它建成了一个不仅在军事上，而且在经济上都具有重要意义的、多功能的基础设施。它不仅减轻了水患，而且还为军事和商用船只提供新的水上通道。更为重要的是，它能够灌溉大量的耕地。这一工程的主要部分包括使用竹笼装卵石，用毛石建造人工岛——它创造出被称为“鱼嘴”的分水功能。这一技术奇迹使岷江为这一地域提供了食物生产的源泉，为这一地域的古中国的百姓孕育文明提供了保障。在古中国的有神论者心中，岷江是神的化身，并由此形成了许多风俗习惯，大量的相关民间传说至今流传。

①作者：Mary Padua，香港大学建筑学院；翻译：刘君。本文主要内容发表在：中国园林，2004，第7期；配图说明：俞孔坚

当地出土的汉代石刻，显示稻田、荷和鱼塘的农业景观
The stone motif excavated from Han Dynasty, showing the agricultural landscape of rice paddy, lotus and fish pond

今天的都江堰水利工程，仍在发挥着重要的作用。这个巨大的基础设施建设对世界文明的贡献已被世界公认，并于2000年被联合国教科文组织列入世界文化遗产名录。这一成就已经成为了当地人民每年值得庆祝的节日性事件。都江古堰以及有关她的民间传说、神话和风俗习惯为土人景观的设计作品——都江堰广场，提供了深厚的创作素材和灵感。而今，这一广场已成为都江堰市新的城市公共开放空间和标志性景观。

## 3.2 作品阐述

认识景观的一个关键点是：景观同时包涵两方面的东西，即景象中包含的物体和使这些物体具有意义的内涵[2]。这两个方面的东西，通过景观解释过程被结合在一起，因而两者是互为依存的。在设计的景观中，我们很容易作同样的论断，景观通过设计师即景观设计师，创造景观并赋予涵义。正像许多人指出的，景观既是场地(site)也是景象——既是“被看的对象”，也是“看法”。景观解释需要有意识地在与日常生活保持距离的状态下观测景观。所以，景观不同于场所（place），这一点非常重要。场所是体验的，而景观是解释的。

景观是随处可见的，我们可以系统地解释我们所见到的景观，汲取很多重要的知识。景观的解释是有关联性（与特殊地域的关联）和条件性的（是谁在解释，为什么这样解释？）[3]。都江堰广场的设计案例中，土人景观对地理区域内的文化层面进行了特定的解释，“借用”当地的历史记忆，运用“隐喻”的设计手法完成了这一作品。Spiro Kostof提醒我们：建筑是脱离不了“背景”和“仪式”的[4]。同样，设计的景观也是离不开“背景”和“仪式”。

Peirce Lewis的经典性著作《阅读景观的原理》 和他关于文化景观的论点在土人景观阅读和解释当地文化的努力中得到了很好的诠释。运用Peirce Lewis的原理要阅读并注意理解与本文相关的如下几点[5]：

(1)景观作为文化的线索

(2)文化统一性和景观等式

(3)普通事物

(4)历史

(5)地理／生态

(6)环境控制

(7)景观的隐喻性

“景观阅读”已经发展成为理解上述文化景观方法和论点的基础教义。Lewis的原理是针对文化景观提出的，一些评论家认为设计的景观也同样符合这一原理。此外，从J.B. Jackson的文章中还可以学到很多有价值的经验，细节请参阅《发现乡土景观》[6]一书。

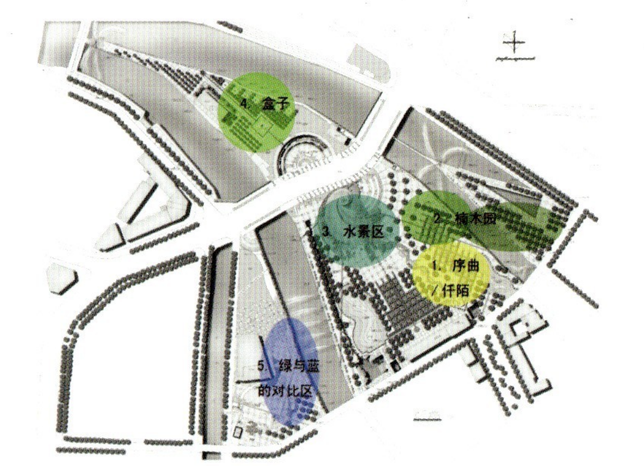


Mary Pudua的都江堰广场空间解释图
Mary Pudua's spatial interpretation

广场中心和几何式的水池设计
The focal sculpture and the grid pattern

从土人景观的设计说明中很清晰地知道都江堰广场的设计灵感来源于当地的都江古堰。而且，土人景观也没有忘记借助当地的农业遗产为其作品中的水系、雕塑设计产生相应的灵感。参考“竹笼”在都江古堰中的使用，土人景观在其作品中甚至创造了新的设计语言。这座城市本身具有的社会和文化肌理在土人的设计作品中得到了深刻体现。这些借鉴于不同层面的元素在整体作品中的有机使用，使其成为独特的富于内涵的设计景观。

## 3.3 评论分析

都江堰广场取代旧有的脏乱之地，成为都江堰市新的标志性景观，她占地$11hm^2$，当地政府部门的目的是使她成为城市再生计划的一部分，政府对她的其余要求可以概括为：

- 城市中心地带景观的提升；
- 供当地居民使用的公共空间；
- 与世界文化遗产都江古堰建立有机联系；
- 成为旅游景点。

土人景观的设计方案在1999年的国际设计竞赛中脱颖而出。他们设计方法源于他们的设计哲学，通过地域、场所以及众多相关联系中寻找设计线索。特别是，土人景观从其掌握的素材中发现并捕捉到了这块土地的本质并因此讲述了一个故事——创作出了一个用当代景观语言讲述都江古堰、地域历史、当地百姓和民间传说的现代人文景观。

空间上，场地本身受到以下诸多因素的制约：

- 主要的车流道路交叉分割地块；
- 穿越场地的水渠的位置和形态是不准改动的；
- 现有水渠将场地分割得支离破碎。

土人景观把这些制约因素视作设计的机遇，着重探索如何通过设计使场地有机地联系在一起，从而产生新的场所个性——“场所性”。在这个设计中，场地被定性为公众开放空间，这是当地政府的要求。

土人景观通过建造下穿式通道，解决了被城市主干路分割的南北场地的连通问题。创造亲水空间、利用邻近水渠的竖向高差设计了水系通道；为增强亲水性，土人景观在广场中引入了溪流。为弱化广场被水渠强行分割的视觉效果，土人景观在设计策略上采用新象征语言，通过雕塑和水平元素的设计，建立广场的整体性。

土人景观在“中心”或“视觉焦点”处设计了规则的几何式景观。设计作品的中心是一座30m高的石雕水塔，其意义是唤起当地民间传说和神话中对岷江水神的记忆，并因此形成视觉地标。同时，土人景观借用几何型广场的设计，让人回忆起“竹笼”和毛石在都江古堰中的应用。一条竖向轴线，从中心雕塑一直延伸到广场的南边界，它由30m的主雕、三个较短的塔和一条线性石廊(导水渡槽)组成。土人景观引入了一条蜿蜒的小溪，缠绕在艺术化的导水渡槽的脚下，参观者可以与水亲近互动。

从这个设计作品的物质属性来看，土人景观为当地创造出了一个新的景观形态。在这个巨大尺度的公众开放广场中，通过各种不同的“户外房屋”和亚空间，形成五个各有特色的区域。空间的多样性和作为场地边界的江水的波涛声，是这个广场给人的整体印象。贯穿整个场地的水的设计，是该作品中最突出的元素，也是广场最与众不同之处。这个设计作品让人回想起劳伦斯·哈尔普林(Lawrence Halprin，1916– )，20世纪加利福尼亚的景观设计师，他在许多美国西部的设计作品里对水的使用，比如华盛顿州西雅图市的高速路公园、俄勒冈州波特兰市的Lovejoy喷泉以及加利福尼亚州圣弗朗西斯科(旧金山)市的Levi Strauss广场等等[7]。

乡土川西平原景观是设计的灵感源泉之一
The local agricultural landscape is one of the main source of the design inspiration

竹笼,独特的治水工具，是艺术设计的灵感来源之一
The Bamboo baskets, the unique hydro technique, is one of the main inspirations of design form

### 3.3.1 五个区域

为了能更深入地分析，我把土人景观设计中的这五个区域——也是我亲自参观体验过的——一一列举说明：

(1)序曲／阡陌
(2)楠木园
(3)水景区
(4)盒子
(5)绿与蓝的对比区

(1) 序曲／阡陌

作为从市内进入广场的南门户和主要入口之一，“序曲”或称为进入广场的南入口部分，为这个新的广场中随后经历的一系列步行空间和场景的展开设定了一个舞台。就像音乐中的音阶一样，这个公共入口空间起到了引导整体设计中其他空间的作用。这是一个安静的前院，由一些方块状的、令人心旷神怡的绿地构成，从某些角度看过去，这块几何化的方格网绿地使人产生比实际更大面积绿地的印象，甚至可以联想到附近的农田。在你的视觉画面中，恒定不变的是场地中心高耸的石塔主雕。在这个令人激动不已的作品中，这个序曲空间尺度不算大，但它却起到了作为水景区的起点和指向作用。

都江堰广场的阡陌景观
The designed agricultural field pattern on the square

(2) 楠木园

我所以称之为楠木园是因为楠木树整齐排列的种植方式，它使人联想到附近农田中整齐排列的果树。这个地块的特性也会使人想到它所在地域的农业传统。在楠木园中有一系列丰富多样的观看河水的角度。沿着楠木园的一侧放眼

郊区的楠木林是设计的灵感源泉之一
The Camphor forest in the suburb is an inspiration to the design

都江堰广场上的楠木园
The Camphor orchard on the square

岷江水系中的卵石和溪流是广场水景设计的构思源泉之一
The water and stone of the Mingjiang River is one of the inspirations for the water feature design

都江堰广场上的水景局部
The pond on the square

望去，视觉焦点处的雕塑凸显于场地的中心。楠木园的“边界”是由导水渡槽的石质漏墙定义的，漏墙从南边的地平，斜向向北延伸，终结于30m高的石制水塔。

漏墙是步行者在不同地块间穿行的“屏风”，也是不同空间之间的过渡。漏墙这种形式的设计灵感来源于都江堰大坝的竹笼的编织语言。楠木园靠近城市一侧则是城市的肌理，北部边缘的一侧就是与主雕和水景相结合的下沉式广场。

（3）水景区

都江堰广场的高潮景观是坐落在中央位置的水塔。它是一座30m高的规则式雕塑，漏刻的斜向网格肌理，象征都江堰水利工程中用来装卵石的竹笼，或许可抽象点说，它好像在回应着从都江古堰传来的水声。雕塑下水池中的红色卵石让人想起岷江中的河卵石。对广场的使用者——人民大众来说，他们能否理解这个设计作品中的象征意义？或者说作品本身能否使大众对多种水元素使用的设计进行思考并因此陶醉于当地的历史？白天，各时段的广场使用者证明了设计作品的成功，甚至在十月四川阴冷的天气里，广场仍然有大量的人流；晚上，许多人们在广场中散步，并把中央水景雕塑（水塔）作为背景拍照。

中央水塔中流出的水汇入小溪沿着线形漏墙蜿蜒流淌。树和其他植物顺小溪边沿生长，产生与主体雕塑周围的硬地景观完全不同的景观体验。视觉焦点景观南面是一组网格状喷泉，它们有涌泉、雾喷等，早晨它们为太极拳的晨练者创造出一幅超现实的奇妙的场景。未来，围绕广场，将有大量建筑以这个广场为前景兴建，这将使广场更加夺目动人。同样，一旦将来其周围出现新的建筑物，下沉式水广场的设计意义将会越来越多地得到体现。

（4）盒子

广场东北部桂花林和林下的多个围合空间是整个广场的另一个兴趣点，它为人们观赏水景、集会活动或即兴表演提供了场所。那些方形的围合空间，为三五成群的人们耍牌和游戏，甚至野餐提供了理想之地。向前走是用树丛和巨石构成的有趣的小型私密空间，这似乎为人们追忆那个古老的水利灌溉工程提供了一个场所。我觉得这里特别有意思，因为这里能听到两侧河水的浪涛声，观赏水的汹涌澎湃，这些巨石正来源于它们之中，这真是一个奇妙的空间。大树提供了充足的荫凉，让人体验到在大自然中所能够得到的享受。

（5）绿与蓝的对比区

位于广场的西南部，“绿与蓝的对比”区域实际上是个隐喻，比喻的是农业和都市生活之间的对比。大面积的绿地与附近水流的对比，当它被视作农业的象征符号时，它和附近包括露天舞台、土人景观的金色天幔和更多城市化硬质景观形成强烈对比。城市化部分的区域经常被用作太极拳的场所和其他娱乐形式开展的地方。公园将有机地融入到城市肌体之中，从这个意义来说，这一区域的设计是成功的。

我所描述的每一个区域，都有多样化的空间，有充足的面积供人们散步和聚会；在设计的尺度方面，人本身是参照体。对软景观和硬景观的区域进行系统设计，合理分布。水元素被进行多样化运用，丰富多彩的水景观设计贯穿其中：包括，几何水池、流线形的小溪和各种喷泉。附近河流中的水浪声在各种不同地块之间形成了有趣的听觉联系和背景，土人景观在其细致的设计中能使这种感觉获得最大的体验。本设计尊重和学习当地都江古堰工程的精神，成功地把握了如此大尺度的壮观景观设计。项目所使用的材料是耐用和持久的。这个设计场所正在被当地的大众使用着。作为都江堰市区复兴的一个项目，时间将告诉我们这个设计作品的成功故事。

设计之初原场地上休闲的人们
The site before design: small places for a small group

### 3.3.2 景观设计中水的应用

在全球背景下，有必要回顾一下水作为一种景观设计元素在历史上和在现代景观中被使用的情况。在西方文艺复兴时期，意大利人是水景设计的大师，经典的例子是 Villa d'Este，这里水被用来创造音乐旋律，各种形式的喷泉——雾喷、拱喷等等。在中国江南文人山水园林设计中，水作为背景使用，总是以静态形式出现。在历史上著名的克什米尔（Kashmir）的穆斯林花园（Moghul）设计中，水被用作运输模式设计，水流作为运输方式，成为花园和宗教旅行的一部分，沿水流分布一系列的活动和事件：泊岸亭、就餐亭、睡眠亭等等。在西欧和美国19世纪的公园运动中，水主要是作为自然的一个组成部分被用作自然风景园林中的湖泊和池塘。

20世纪60年代后，公共空间被作为城市肌体的一个部分来设计。这些公共空间成为有世界声誉的加利福尼亚景观设计师Lawrence Halprin的“画布”，他职业生涯中获得过各种奖励。他对水景观的创作灵感来源于自然。他建造的公园和公共空间通过水及其与自然关系中获得灵感。其经典作品包括Lovejoy公园、西雅图高速路公园和富兰克林·罗斯福广场。在美国景观设计领域，他可能是第一个在景观中重新发现并引入水元素作为创作主题的人。景观设计事务所，包括SWA，EDAW和其他的后来者们都在从他的创作成果中吸取营养。

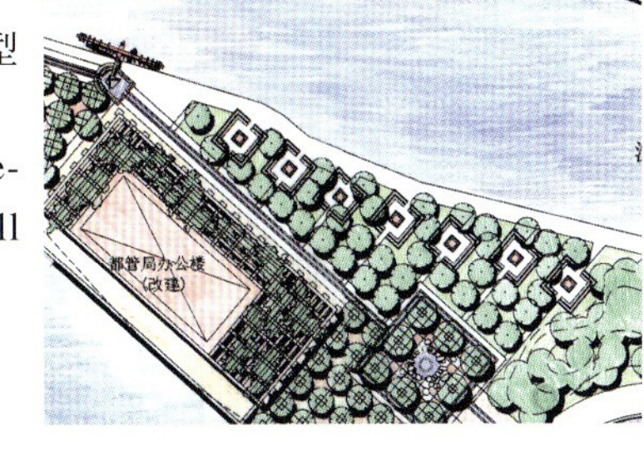

盒子：设计后的微型休闲空间
The resting box: designed for the small group activities

### 3.3.3 场所的营造和城市复兴

土人景观对水元素的使用，是对水的新解释，为中国当代景观设计探索了新的设计语言。这种水的设计方法与中国古典园林中的水设计是根本不同的。水作为主导景观元素，其设计灵感来自都江古堰的启示。Lawrence Halprin从当地的High Sierra Mountains山区获得灵感，这种灵感仅仅用于跌泉的设计。他在公园和广场的其余设计中都是将水作为景观焦点来使用的。他的城市公园广场设计与当地的城市肌理是紧密联系的，但并没有浸透当地的文化内涵，然而，土人景观的设计却深刻地蕴涵了当地的文化。

绿与蓝的对比区
Contrast of green and blue

都江堰广场的设计被地域场所的文化气息和乡土气息所强化。这个设计作品与邻近河渠中奔腾的水流、浪涛声和各种设计元素有机地融为一起。水元素的引入和所有土人景观引入的设计形式：雾喷泉、主雕塑、小溪、下沉式水广场等等，构成一部交响乐，讴歌着都江古堰的水利盛事。

土人景观设计中水元素的使用和当地现有的关联背景使这个新的城市公共空间具有强烈的个性。这个最新创造的个性，昭示了作为景观设计师和城市设计师的一个重要职业重任：场所的营造。这个新景观为城市的建筑物和都市风貌提供了一个独特的景观背景，并大大增强了城市本身的特色。都江堰广场的战略地位为城市再生提供机会，并在使都江堰成为国际旅游目的地的努力中发挥重要作用。

主体雕塑:投玉入波
The focal sculpture：a jade dedicated to the wave

## 注解

[1] See Chapter 5, Ancient Sichuan and the Unification of China, by Steven F. Sage for a comprehensive discussion of the Dujiangyan inception and relationship to land reform, endeavors in war and agriculture.
[2] See Cosgrove 1985; Mitchell 1996; Corner 1999
[3] See Meyer 1997
[4] See Kostof, 1985
[5] See Lewis
[6] See Jackson 1984
[7] See Chronology in Lawrence Halprin: Changing Places

## 参考文献

Baker, Alan R.H. 1992. Introduction: On Ideology and Landscape. In Alan R. H. Baker and Gideon Biger, Eds, Ideology and landscape in Historical Perspective. Cambridge: Cambridge University Press

Corner, James (ed.). 1999. Recovering Landscape: Essays in Contemporary Landscape Architecture. Princeton: Princeton Architectural Press

Cosgrove, Denis. 1985. Prospect, Perspective and the Evolution of the Landscape Idea. Transactions of the Institute of British Geographers, 10:45~67

金色天幔夜景
The golden canopy in the light


Fried, Helene. 1986. Lawrence Halprin: Changing Places. San Francisco Museum of Art, San Francisco, California

Golley, Frank B. 1998. A Primer for Environmental Literacy.  New Haven: Yale University Press

Jackson, J. B. 1984. Discovering the Vernacular Landscape. New Haven: Yale University Press

Kostof, Spiro, A History of Architecture: Settings and Rituals. New York: Oxford University Press

Lewis, Peirce. 1979. Axioms for Reading the Landscape: Some Guides to the American Scene. In The Interpretation of Ordinary Landscapes, edited by Donald Meinig, Oxford: Oxford University Press

Meyer, Elizabeth K. 1997. The Expanded Field of Landscape Architecture. In Ecological Design and Planning, edited by George F. Thompson and Frederick R. Steiner. New York: John Wiley and Sons

Mitchell, W.J.T. (ed) 1994. Landscape and Power. Chicago: University of Chicago Press


## 作者介绍


Mary Padua为香港大学建筑学院助教(assistant professor)。在成为全职教授之前，曾在美国加利福尼亚、香港的政府部门和私人事务所从事景观设计和管理达二十多年。1978年获加利福尼亚大学伯克利分校景观设计学专业学士学位，1984年获加利福尼亚大学洛杉矶分校建筑学与城市设计硕士学位。研究兴趣包括当代景观设计和城市环境中的乡土景观。

# 1 Dujiangyan Square as people Places

Name of Project: Dujiangyan Square
Project Location: Dujiangyan City/Sichuan, China
Project Type: Urban square, public space in central city
Size: 11 hectares
Date of Completion: The 1st of May, 2003
Client: Dujiangyan City Government

Dujiangyan Square is located in the middle of a dilapidated and featureless old townscape of Dujiangyan City, in Chengdu,Sichuan Province, China. In 1999, an international competition was held. Ten entries were selected for the short list and the final selected design was executed by the 1st of May, 2003.

At the fact that the project is located in the center of a densely populated Chinese city, where open space is badly needed for the normal people, this design was able to provide diverse spaces and places for the old, the young, the healthy and the disable, and for play, exercise, meeting and watching. This design was inspired by the unique regional natural and cultural landscapes, irrigation works, and local living styles. With a budget of less than 40 US dollars per square meters, the project has created an artful urban space that tells ancient stories in a modern language, expresses the regional and local identity yet in a new approach, and was designed to attract tourists as well as accommodate the daily needs of the local citizens.

## 1.1 The region and the site

In a legendary basin surrounded by high mountains, Chengdu is agriculturally one of the most productive and romantic places described in Chinese, "country of the heaven", and is one of the most densely inhabited areas. The city Dujiangyan was historically named as the Irrigation County that was named after the famous ancient irrigation works, the Dujianyan Weir. The prosperity of Chengdu Basin has been dependent on this historic infrastructure project. The weir, a world cultural heritage, was built more than two thousands years ago, and is still in use today.

The site was a former derelict urban area, and was totally cleared and eliminated before the design competition was held. It is about two kilometers away from the famous Dujiangyan Weir. The canal diverted from the weir is divided into four irrigation canals that run through this site.

导水漏墙
The carved stone wall and aqueduct

## 1.2 Project purpose and intents

The purposes of the project are: to provide diverse open space in the middle of the densely populated city, for the residents to play, exercise, meet and enjoy and improve the landscape of the downtown area to recreate a sense of place identity of the Dujianyan City.

主体雕塑:投玉入波
The focal sculpture：a jade dedicated to the wave

## 1.3 Challenges and solutions

Being in the center of a busy and featureless urban matrix, the design faced a lot of challenges, including:

(1)The massiveness of the people and their diverse needs for open spaces: These needs include morning exercises for the working class, play spaces for the kids (there is not a single playground in the whole city!), a getting together place for cultural and entertaining activities, and places for the increasing number of old residents who enjoy sitting in groups playing cards and watching birds. Design solutions include special and temporal differentiations, and multiple use of spaces.

(2)Transportation: A main street across the existing site cuts the square into the south and north parts, the design therefore created an underpass in association with sunken water squares at both ends so that the two separate parts can be connected functionally and spatially.

(3)Water accessibility: The slope of the four irrigation canals are steep with rapid water flows, and are not allowed to be modified due to water management regulation. The design made use of the elevation difference to divert water using traditional ways from the upper reach of the canals and create a creek in the square, allowing the water to become accessible and touchable.

(4)Fragmentation and identity: The square was fragmented by rapid canals. A formal and symbolic language was developed at the inspirations of the bamboo basket as a unifying element, which gives identity and uniformity to this public space. The overall layout of the square resembles an unfolding bamboo basket, that radiates from a focal point, at which stands a sculpture. Also an abstraction of a bamboo basket are the lighting columns, the carved stone walls, the fences and the detailed paving. All of these elements are intended to use similar visual language.

## 1.4 The inspirations for the design

In addition to formulating strategies that make the square function as places for the people, artistic and aesthetic issues and symbolism are key considerations in the overall design approach. The designers took their design cues and inspirations from the following research and experiences.

**(1)The living style of the people:** Dujiangyan City is well known for its leisurely living style, which was also a contribution from the reliable irrigation system that has turned the basin a productive and worry-free "land of heaven." The way people meet in groups, play cards in groups and site in line in the sun, work and stay in family, all become inspirations for the site design.

**(2)The irrigation works:** The legend of Li Bing, the inventor of the Dujiangyan Weir: Li Bing tried in vain many kinds of techniques to build the strongest weir against floods. He tossed treasure stones into the waves as tributes to the river goddess for advice. He was inspired to use the softest devices against the strongest forces, which turned out to be bamboo baskets stuffed with pebbles, and the wooden poles in triplet. For two thousands years, these soft and simple weir-making techniques have been to be the most effective, economic and durable, stronger than steel and concrete. And this technique is still in use today. Other techniques of irrigation works such as fish mouths (triangle-shaped dividers) and flumes, etc. are also inspiring for visual language for the design and the art works.

**(3)The landscapes of the region:** The agricultural landscapes, the field patterns associated with the irrigation system, especially the unique stretches of rapeseed in blossoms. The local vernacular architecture that uses wooden frame structures, and the courtyard with stone walls, etc. are all inspiring visual forms for the design.

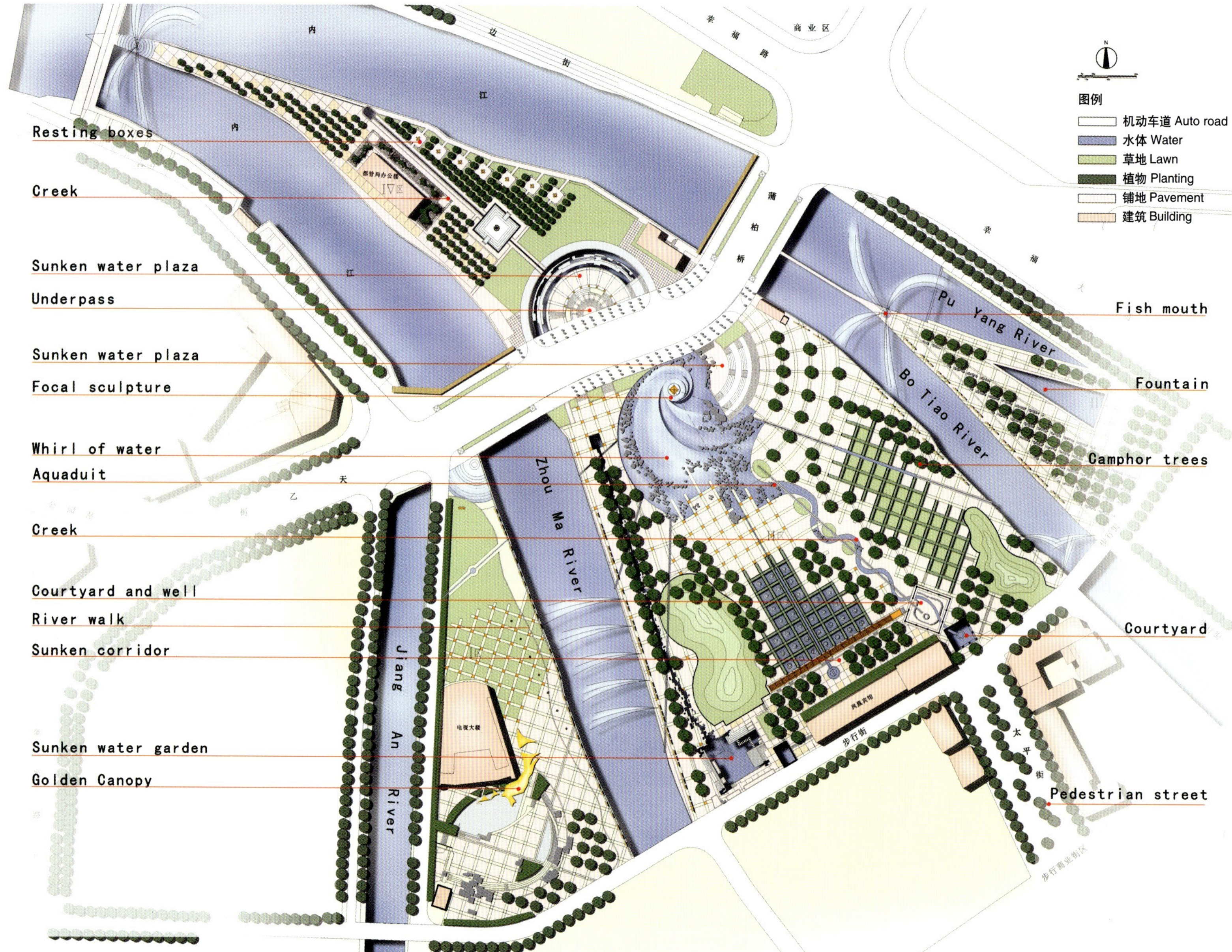


The master plan

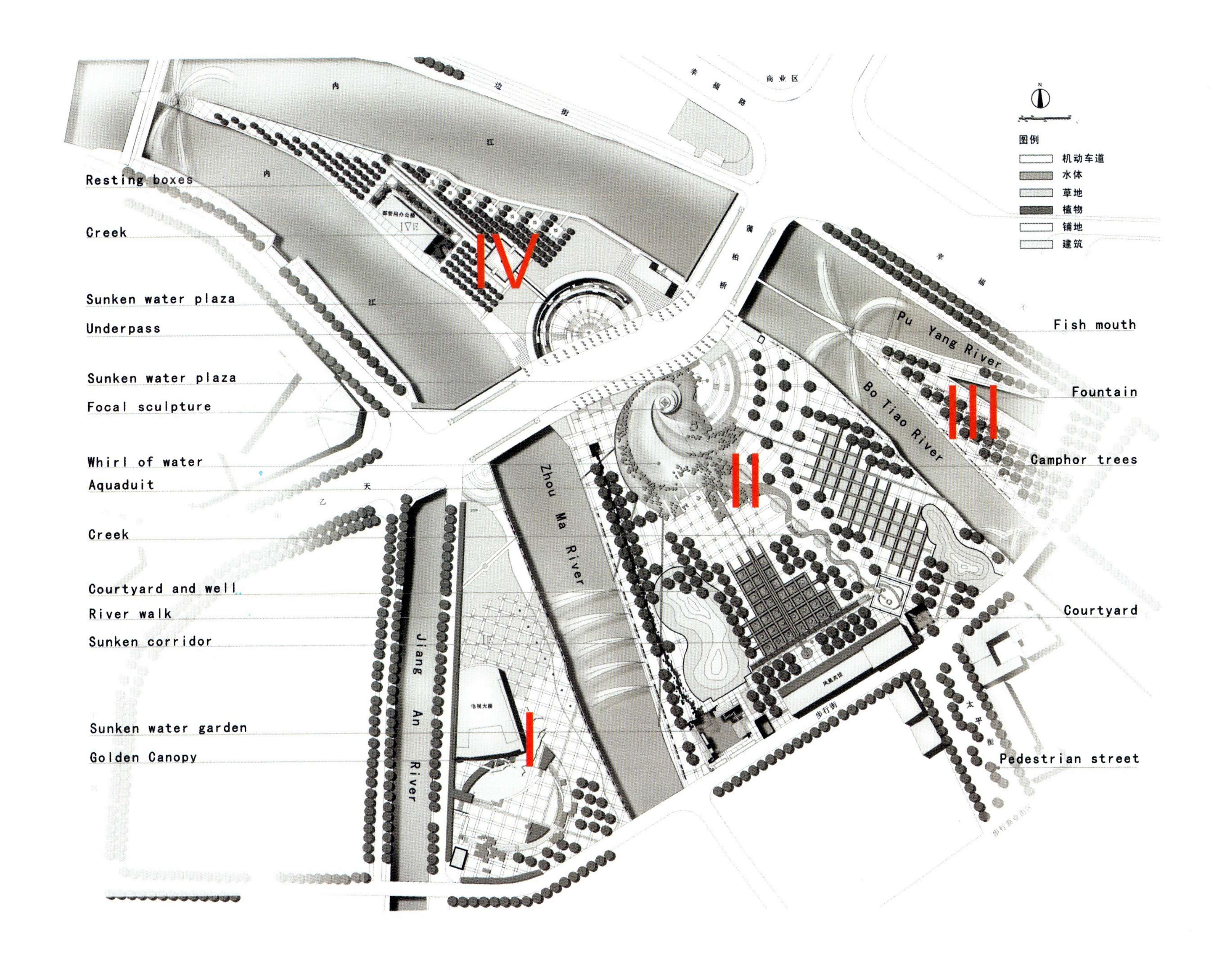
图例
机动车道
水体
草地
植物
铺地
建筑
Resting boxes
Creek
Sunken water plaza
Underpass
Sunken water plaza
Focal sculpture
Whirl of water
Aquaduit
Creek
Courtyard and well
River walk
Sunken corridor
Sunken water garden
Golden Canopy
Fish mouth
Fountain
Camphor trees
Courtyard
Pedestrian street
Pu Yang River
Bo Tiao River
Zhou Ma River
Jiang An River
IV
III
II
I
商业区
步行街

Four phases

## 2 Some Important Features of the Design

### 2.1 The layout

In concert with the site topography, the plan is in a radiating pattern with the focal point controlled by the main sculpture, and a system of various water features were integrated using elevation, allowing the water features to be accessible numerous native camphor trees were planted to provide shade, and the fragmented pieces were formally unified with a visual language abstracted from the bamboo basket.

Divided by the four running through rivers, the square is naturally composed of four sub-areas or phases:

Phase-I: at the south west side of the square, featured with an amphitheater, the Golden Canopy and a river walk.

Phase-II: the central part of the square, featured with the focal sculpture, the cascade, the carved stone wall, sunken water garden, etc.

Phase-III: at the south east side of the square, featured with a big fountain.

Phase-IV: at the north end of the square, featured with small boxes in the forest of sweet scented osmanthus.

### 2.2 People's places

Various spaces are allocated in different parts of the squares, they are designed with the culture of the local people in mind, including seating boxes of $5 \times 5$ meters designed for groups to play cards. A sunken amphitheater, and three sunken water spaces are designed for people to enjoy water features. The fountains are designed for people to play in, and seats on the sides of the canals where they used to be, etc.

### 2.3 The central axis

It is a combination of the focal sculpture of "a piece of jade contributed to the river goddess" , three lighting columns, a carved stone wall of about 100 meters long with water flow on the top, and a meandering creek that diverted from the upper reach of the canal flowing along the axis. This axis along a diagonal line that visually connects the pedestrian street at the south end and points to the valley at the north kilometers away, where the famous ancient Weir is located. The focal sculpture is a 30 meters high and 3 meters in width, in granite. At its base is a whirl shaped water feature, a combination of fountain and cascade. The carved stonewall is used to create spatial diversity, strengthen the central axis, and deliver the meanings associated with the bamboo fences and carved brick walls in Chinese architecture, and also flumes widely used in this region.

### 2.4 Art works

Some of the important art works include the metal Gold Canopy, hung on tilted bronze poles which collectively assembe the wooden triplets used in the ancient irrigation works. The golden canopy also recalls the experience of the stretches of rapeseed blossoms.

It is a public space for all people and diverse activities, that tells the stories, the stories of the place, the story of the past, and the story of the common people.

网纹水波

The waved cascade

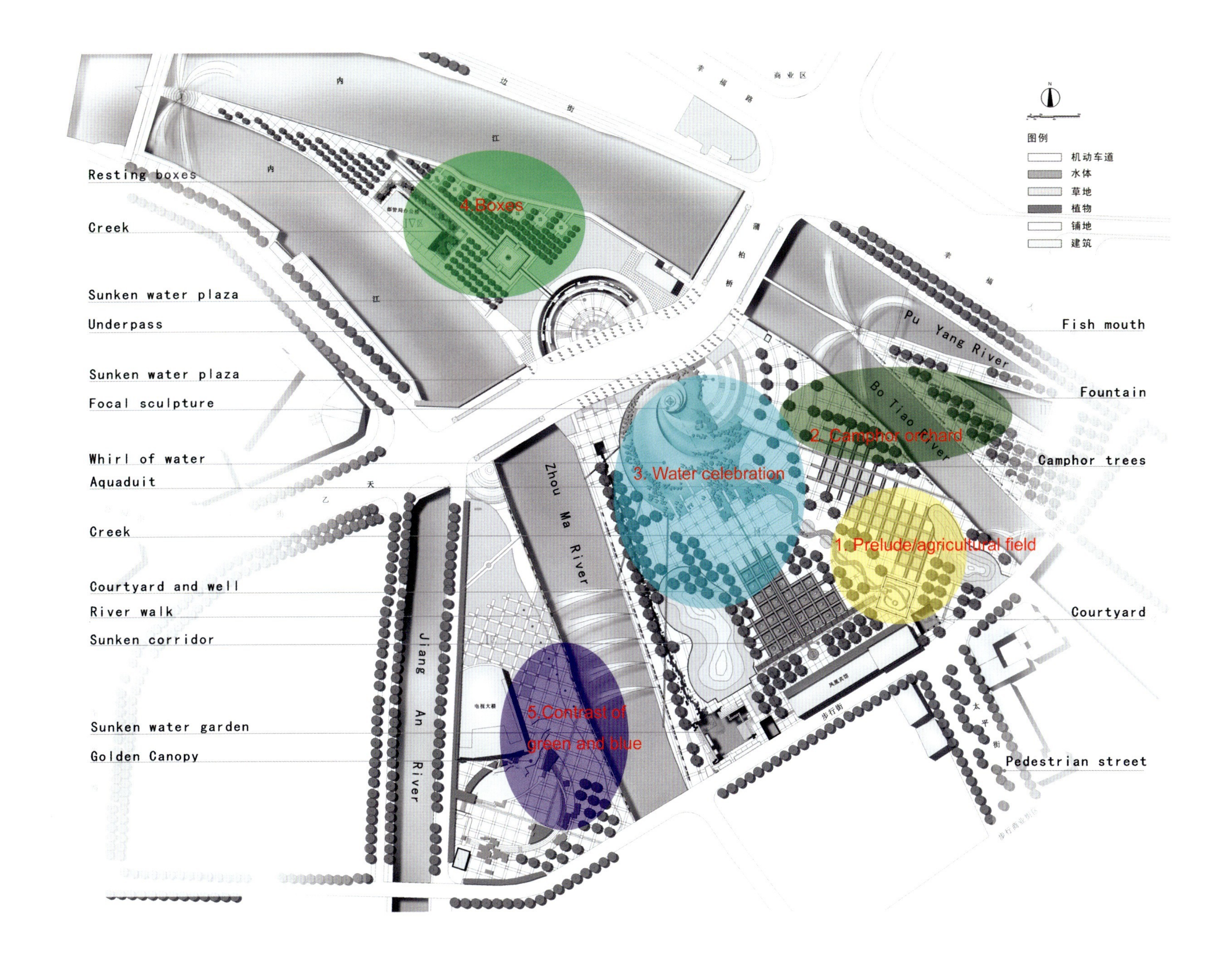


Mary Pudua's spatial interpretation of the square

# 3 Critique: Dujiangyan Square-Articulating a narrative public open space ——

***Homage to the nearby 2400 year-old Dujiangyan Irrigation works***

By Mary Padua, ASLA, CLARB, RLA
Faculty of Architecture,University of Hong Kong

Turenscape tells an ancient story by creating a narrative landscape that echoes the neighboring World Heritage site Dujiangyan Irrigation Works, and the local culture that helped to create it. As a critic for the project, I am curious about the place of the project. Where is this landscape? What is adjacent to it? What previously existed on the site? I am also curious about the temporal aspects of the project. How will the project change over time?

To give context and critical analysis for the project, Dujiangyan Square, I have organized the following paper to include a brief cultural and historical background of the site and area, a discussion of narrative and metaphor, project review and analysis, the use of water as a design element in contemporary landscape architecture, and a discussion of the project within the framework of place-making and urban regeneration.

## 3.1 Cultural / historical context

To understand the creation of Turenscape's Dujiangyan Square, it is important to briefly describe the area's historical and cultural context. Dujiangyan, formerly known as Guanxian, is located 60 km northwest of Chengdu, the provincial capital of Sichuan. Sichuan province has been established as culturally significant and as a major cradle of Chinese civilization in ancient times. It is also known as "heavenly kingdom" due primarily to its rich agricultural resources that are a result of the ancient irrigation works. This rich history provides a strong foundation for modern Sichuan's identity, and the city of Dujiangyan.

More importantly from a historical perspective, during the Qin period, land reform led to major public works that would allow the transformation of dry land into productive agricultural areas to thrive and feed armies. This was part of a larger strategy to unite ancient China. Dujiangyan in ancient Chinese translates as "Capital River Dam". It was a colossal, carefully planned public works project in the eastern half of the Eurasian continent, the first known in the middle third century B.C.[1]

This engineering project was established to harness the Min River, one of the Yangzi River tributaries. The project built under the provincial military governor, Li Bing, was a multi-faceted infrastructure project with both military and economic claims. It was intended to alleviate a chronic flood danger, provide a new inland waterway for commercial and naval boat traffic, and more importantly irrigate agricultural fields over a vast area. Building this major

浅水池
The shallow pond with red sand stones

infrastructure project involved the use of river rocks contained in bamboo baskets, and an artificial island made of piled stone to create a functional feature known as the "fish bite". This technical wonder and the Min River also provided major sources of food to the area, and brought with it the life and culture of the ancient Chinese local population. Ancient animist Shu religion had regarded the Min River as a deity. Various ancient local customs and beliefs were formed around the sacred aspects of the Min River and created folklore that is still believed today.

Currently, the Dujiangyan Irrigation Works continues to operate and function. Its significance as a major infrastructure project was recognized as a cultural contribution worldwide and was designated a World Heritage site in 2000 by UNESCO. It is currently the site of various local and regional festivals. This ancient infrastructure project, its folklore and mythology, and local tradition provide the inspiration for Turenscape's design of a new public plaza as a major focal point and destination for the city of Dujiangyan.

## 3.2 The narrative and metaphor

It is critically important to understand that landscape is two things at once: a collection of material objects within a scene and the ideas that make those objects meaningful.[2] The two are brought together in the act of interpretation and are therefore mutually constituted. In the case of the "designed landscape", we can easily formulate the same statement and make clear connections about a landscape that is meaningful and created by a designer—in this case the landscape architect. As many have pointed out, landscape is both site and sight — both "what is seen" and a "way of seeing." Landscape interpretation requires a deliberate act of looking by a certain distancing for the spaces of everyday life. So it is also critically important to distinguish "landscape" from "place". Places are experiences; landscape are interpreted.

Landscape is everywhere and we can learn important lessons by interpreting what we see in a methodical manner. Landscape intrepetation is contextual (linked to specific site) and situational (it asks: who's interpreting and why?).[3] In the case of the designed landscape, Dujiangyan Square, Turenscape has made specific intrepetations of the cultural aspects of the geographical area, using metaphors for design, and "borrowing" from the memories of the area. Spiro Kostof reminds us that architecture is inextricably bound in both "settings" and "rituals".[4] Similarly, the *designed landscape*, also falls within these boundaries.

Peirce Lewis' seminal work, *Axioms for Reading the Landscape,* and his notions about cultural landscape apply in Turenscape's efforts to read and interpret the landscape of this region. Peirce Lewis axioms while worth noting should be read within the context of the essay it was written for[5]:

(1) Landscape as clue to culture

(2) Cultural unity and landscape equality

(3) Common things

日常人的空间
The everyday's place

(4) History
(5) Geography/ecology
(6) Environmental control
(7) Landscape obscurity

"Landscape reading" has grown as a basis for teaching this method and notions of cultural landscapes. While this method is about cultural landscapes, some critics now believe that *designed landscapes* may meet the test for Lewis's axioms. Furthermore, other valuable lessons could be learned from essays written by J.B. Jackson, in particular, *Discovering the Vernacular Landscape*.[6]

It is clear from Turenscape's description of the project that the design of Dujiangyan Square was inspired by the nearby ancient infrastructure project. Turenscape also borrows from the agricultural heritage of the region in its use of plant materials and for inspiration of the water features and sculptures. References to the bamboo basket are used to create a new design vocabulary for the project. The social and cultural fabric of the city itself is also incorporated in the design. Borrowed elements from several of these layers are incorporated into the overall design and contribute to this unique narrative landscape.

## 3.3 Critical analysis

Dujiangyan Square replaces an old park in a derelict area of Dujiangyan. It covers eleven hectares and the objective of the local municipal authorities was to regenerate this part of the city. Other local government requirements called for:

- the improvement of downtown's landscape
- provision of public open space for the local residents
- creating a link to the nearby World Heritage site, Dujiangyan Irrigation work
- creating a tourist attraction and destination

Turenscape was selected after an international design competition held in 1999. Their design approach was based on their philosophy of landscape architecture to extract design clues from the site, the area, and the many issues affecting the site. In particular, Turenscape saw its charge to find and capture the essence of the place and to tell a story — create a contemporary narrative landscape that portrays the history of the nearby Dujiangyan Irrigation Works, the region, its people, and folklore.

Physically, the project was constrained by many elements:

- major vehicular corridor that crosses and separates the site
- the structure and location of the water canals were fixed and unchangeable
- these existing water canals segregated and fragmented the site

灯柱
The lighting post

Turenscape viewed these constraining elements as design opportunities, and explored ways that would help weave the site together and create a new identity and "sense of place". In this case, the place is a public square that was designed to meet the requirements of the local municipal authorities.

Turenscape tackled the constraining road by building an underpass that would unite the main northern and southern parts of the project. To create water access and take advantage of the vertical elevational differences of the adjacent canals, Turenscape introduced a creek as a water feature in the plaza. To mitigate against the segregating character of the canals, Turenscape devised a design strategy that imposed new formal and symbolic language in the form of sculpture and horizontal elements that would unify the project.

Turenscape utilizes an ordered and structured geometry that is based on a "center" or focal point. In this case, the center is a 30-meter high water tower carved in stone, and its design is meant to evoke the local folklore and mythology of the Min River as goddess, and provide a visual landmark for the area. At the same time, Turenscape invokes a geometric plaza design that symbolizes an unfolding bamboo basket. This design vocabulary is reference to the bamboo baskets and rocks that were used to build the nearby ancient Dujiangyan irrigation works. Emanating from the central water tower is a series of three shorter towers and carved linear stonewall (aqueduct) that extends across the plaza to the boundary of the site. Along this dominating rectilinear element, and spouting from the base of the central focal point, Turenscape introduces a curvilinear creek where park visitors are allowed to interact with the water.

By the nature of the physical aspects of the design, Turenscape was able to create a new landscape form for the area. Five discrete precincts are distinguishable in this large-scale public plaza with various "outdoor rooms" or sub-areas within them. The variety of spaces and the sounds from the raging water are an overarching reminder of the moving water that envelops the edges of the site. The use of water throughout the site is the most distinctive design feature and dominating design element in the site. The project is reminiscent of the work of Lawrence Halprin (1916- ), highly 20$^{th}$ century California landscape architect and the way he uses water in several of his projects built in the American West like, Freeway Park in Seattle, Washington, Lovejoy Fountain, Portland, Oregon, and Levi Strauss Plaza in San Francisco, California.[7]

For the sake of further analysis I call the five precincts that I experienced in the project the following:

(1) Prelude/agricultural field
(2) Camphor orchard
(3) Water celebration
(4) Boxes
(5) Contrast of green and blue

***Prelude/agricultural field***

As one of the major gateways and urban entry points to the plaza, the "prelude" or entry plaza sets the stage for the coming series of pedestrian events in this new open space. As in a musical score, this prelude acts as the introduction to the rest of the design composition of the plaza. This forecourt is a quiet inviting space with lawn in geometric forms that when viewed from certain points evoke a larger field of green landscape, reminiscent of the nearby agricultural fields. Constant in your viewing plane is the tall stone tower at the center of the site. The scale of this area is small to medium in scale and provides a starting point and orientation to the celebration of water that unfolds in the experience of this exciting new project.

***Camphor orchard***

I call this camphor orchard because of the ordered way the camphor trees are laid out. It is reminiscent of agricultural orchards and systematic ways that the nearby agricultural fields are laid out. The area is also a reminder of the area's agricultural heritage. Within the camphor orchard there is a variety of sitting areas, and places to view the canals. Along one side of the camphor orchard, the sculptural focal point of the site begins to project itself from the center of the site. The "edge" of this precinct is created by the linear stonewall that emanates from the 30-meter high central water sculpture.

This stonewall is a screen that allows pedestrians to walk through from one area of the site to another, a design device that also creates a transition from one place to another. The style and design vocabulary of the stone screen wall borrows from the woven bamboo baskets used to build the nearby ancient dam. Along the more urban edge of the camphor orchard is the urban fabric of the city. Also, along one edge is the sunken water plaza that is part of the central water feature.

***Water celebration***

The climax of the site is the dominant water tower that is centrally located. It is a formal 30-meter tall sculpture that appears to symbolically celebrate water, and perhaps, abstractly, harkens back to the ancient irrigation works nearby. The red stones in the water pond are reminiscent of the rocks found in the Min River. The penetrating grid of the hardscape is an attempt to remind one of the nearby ancient irrigation works, and the bamboo baskets. For common park users, will they see this symbolic gesture or is it enough for the user to ponder the significance of the various water elements and enjoy the experience of the place? The success of the project is demonstrated by the users of the space on the various times of the day, and there was much evidence of the local population using the park, even during the overcast cool autumn October weather. In the evening, many park strollers would take photos with the central water tower as the background.

The water from the tower is channelized into a creeklike form that weaves under the linear stone screen wall. Trees and plants align edges of the creek creating a different experience from the central focal point and the hardscape around it. South of the focal point is a grid of water fountains that perform different functions, bubbling, misting, etc. It provides a surreal backdrop to the early morning *taichi* practitioners. Ultimately, this area will be more successful as a foreground plaza to future building along this edge of the site. Equally, the sunken water garden would have more design meaning once a new building is in place.

***Boxes***

The boxes area at the northwest of the "water celebration" provides another focal point for a place where people can gather and watch performances, either planned or impromptu. The smaller seating squares in this sub-precinct provide places for groups of people to play games and picnic, a local pastime. As one moves further along, there is an interesting small-scale intimate space with trees and large boulders that are reminiscent of the ancient irrigation works. I found this place to be most intriguing as I could hear the sounds of the adjacent water, see the raging water, and imagine the river where these large boulders came from. The mature trees provide sufficient cover to create the sense of intimacy that some places in nature might give.

***Contrast of green and blue***

The precinct of "contrast of green and blue" is a metaphor and creates a contrast between agriculture and urbanism. The large field of green lawn contrasts with the adjacent water. When viewed as a symbolic gesture of agriculture, it is in contrast with the adjacent hard paved area that contains a stage, and Turenscape's Golden Canopy, a much more urban part of the project. This area is highly used as an impromptu performance area for *tai chi* and other forms of entertainment. As a gesture towards integrating and weaving the park into the urban fabric, this area succeeds.

Within each of the precincts I describe, a variety of places are created. Ample areas for walking and gathering are provided. Water elements are used in various ways and the scale of the project offers locations that are human in scale. A hierarchy of landscaped and hardscaped areas are distributed throughout the project in a systematic way and are integrated by a series of water features: hard-edged channels, curvi-linear creek, and a variety of water fountains. The sound of the adjacent raging waters provide an interesting acoustical context in various areas of the site, and Turenscape was able to maximize this sensual aspect in their sensitivity to design. While the overall project is grand in its scale, it pays homage to the nearby ancient irrigation works. The materials that are used are durable and long lasting. The place is being used by the local population. As a project to regenerate downtown Dujiangyan, the passage of time will tell this part of the success story.

## 3.4 Water use in landscape architecture

To provide a larger global design context, it is critical to point out the way water as a design element has been used in designed landscapes historically, and in contemporary landscape architecture. In the western Renaissance, Italians were the masters of using water as a spectacle and as an animator of designed landscapes. The ultimate example is the Villa d'Este, where water was used to create a musical water organ, and channeled to create various forms of ordered water fountains in different forms — spraying, arching, etc. In the Scholar Gardens of the Jiangnan style, water is used as a backdrop, and is always still, never in motion. In the historic Moghul gardens in the veil of Kashmir, water is used as a transportation mode, is traveled on and is part of the garden and religious journey, where a series of movements and events occur: arrival pavilions, eating and sleeping pavilions, etc. In the $19^{th}$ century park building movement in Western Europe and America, water is primarily an aspect of nature that is often presented as lakes or ponds in romantic parks.

By the 1960's, public spaces were being designed as part of the urban fabric. These public spaces became the canvas for Lawrence Halprin, world-renowned California landscape architect, who has been awarded numerous medals and awards throughout his career. His career was built on the celebration of water inspired by nature. The parks and public places he built were inspired by water and its relationship to nature. Seminal works included Lovejoy plaza, Seattle Freeway Park, and Franklin D. Roosevelt plaza. He was probably the first to re-introduce water as a theme in the field of landscape architecture in America. Corporate landscape architects, SWA, EDAW, and others followed suit in their designs of private corporate landscapes.

In response to the environmental movement of the 1960's & 1970's water became ecologically charged, and later landscape architects became more involved in the design of water treatment plants and wetlands design. By the late 1980's and 1990's, landscape architects returned to ideas of landscape architecture as a fine art, using water as a design element, not in ecological terms. A few landscape architects attempt to weave art and science in their expression of the use of water. George Hargreaves and his attempts to design the Guadalupe River Park corridor is an example of this.[8] Betsy Damon and Margie Ruddick demonstrate how to clean water in the Living Water Garden in nearby Chengdu, allegedly the first of its kind. Although, often the idea of cleaning polluted water is often discussed in design studios in many western universities.[9]

## 3.5 Place-making and urban regeneration

Turenscape's use of water falls into a new design intrepretation of water and creation of new design vocabulary in the field of contemporary landscape architecture in China. It goes against the grain of the water is utilized in the classic Chinese Scholar Gardens. While water is the dominant theme, it is inspired by an ancient public works project, and the heritage and folklore surrounding its creation. Lawrence Halprin took his design cues from the nearby High Sierra Mountains, for water fountain design only. The rest of the design expression of Halprin's water-dominated parks and plazas were intended as focal points. The design of Halprin's urban parks and plazas was responsive to the adjacent urban fabric and did not imbue any overlay of cultural heritage, whereas clearly Turenscape's design approach embraces the local cultural context.

Dujiangyan Square's design is strenghthened by the culture of regional "place" and the vernacular landscape. Its design coupled with the raging torrential water in the adjacent canals bounds the site and its sound along with the various design elements weaves the site together. The introduction of water elements, and all of the forms that Turenscape introduces, field of misting fountains, central sculptural water fountain, creek, sunken water gardens, etc, all are orchestrated as part of a larger new celebration of the ancient Dujiangyan irrigation project.

Turenscape's use of water and the existing context gives an strong identity to this new public square. This newly created identity forges an emphatic notion of place-making, a design principle that many landscape architects and urban designers live by. This new place offers a unique backdrop for the buildings and cityscape and enhances the city's identity. Its strategic location offers this area of the city the potential for urban regeneration, and its ongoing efforts as an international tourism destination.

### *Endnotes*

[1] See Chapter 5, Ancient Sichuan and the Unification of China, by Steven F. Sage for a comprehensive discussion of the Duijiangyan inception and relationship to land reform, endeavors in war and agriculture.

[2] See Cosgrove 1985; Mitchell 1996; Corner 1999

[3] See Meyer 1997

[4] See Kostof, 1985

[5] See Lewis

[6] See Jackson 1984

[7] See Chronology in Lawrence Halprin: Changing Places

[8] See Thompson

[9] Based on author's conversations with studio teachers, Margie Ruddick (Harvard, U. of Pennsylvania), Pamela Burton (Sci-Arc, USC), Margaret Crawford (Harvard), John Kaliski (Sci-Arch, U. of Michigan)

### *References*


Baker, Alan R.H. 1992. Introduction: On Ideology and Landscape. In Alan R. H. Baker and Gideon Biger, Eds, Ideology and landscape in Historical Perspective. Cambridge: Cambridge University Press

Corner, James (ed.). 1999. Recovering Landscape: Essays in Contemporary Landscape Architecture. Princeton: Princeton Architectural Press

Cosgrove, Denis. 1985. Prospect, Perspective and the Evolution of the Landscape Idea. Transactions of the Institute of British Geographers, 10:45~67

Fried, Helene. 1986. Lawrence Halprin: Changing Places. San Francisco Museum of Art, San Francisco, California

Golley, Frank B. 1998. A Primer for Environmental Literacy. New Haven: Yale University Press

Jackson, J. B. 1984. Discovering the Vernacular Landscape. New Haven: Yale University Press

Kostof, Spiro, A History of Architecture: Settings and Rituals. New York: Oxford University Press

Lewis, Peirce. 1979. Axioms for Reading the Landscape: Some Guides to the American Scene. In The Interpretation of Ordinary Landscapes, edited by Donald Meinig, Oxford: Oxford University Press

Meyer, Elizabeth K. 1997. The Expanded Field of Landscape Architecture. In Ecological Design and Planning, edited by George F. Thompson and Frederick R. Steiner. New York: John Wiley and Sons

Mitchell, W.J.T. (ed) 1994. Landscape and Power. Chicago: University of Chicago Press

Thompson, J. William, The Poetics of Stormwater, *Landscape Architecture*, Vol 89, Number 1, Washington: American Society of Landscape Architects, 1996, 132~139, 144, 146


### *Author's Biography*

Mary Padua is Assistant Professor in the Department of Architecture, Faculty of Architecture, University of Hong Kong. Prior to her full-time academic appointment, Ms. Padua practiced for over twenty years in the public and private sectors in California and Hong Kong. She received the Bachelor of Arts in Landscape Architecture from the College of Environmental Design, University of California, Berkeley (1978) and the Master of Arts in Architecture and Urban Design from the Graduate School of Architecture and Urban Planning, UCLA (1984). Ms. Padua's research is focused on contemporary designed landscapes and vernacular landscapes in urban settings.

# 4 图解都江堰广场

## An Illustrative Interpretation

露天演艺广场：圆形广场，包括舞台、金色天幔、竹林分隔的后台排演场地。有高低错落的花岗石铺装构成了一个富有情趣的观演场所

The amphitheater: composed of a theater, a golden canopy sculpture, a bamboo encircled a private practice area. The stepped stones created interesting seating area for the audience

I区总平面图
The plan of phase-I

此图中修改了演艺舞台的弧形看台部分的台阶及残疾人坡道,具体见图中。

露天演艺广场平面图

The plan of the amphitheater

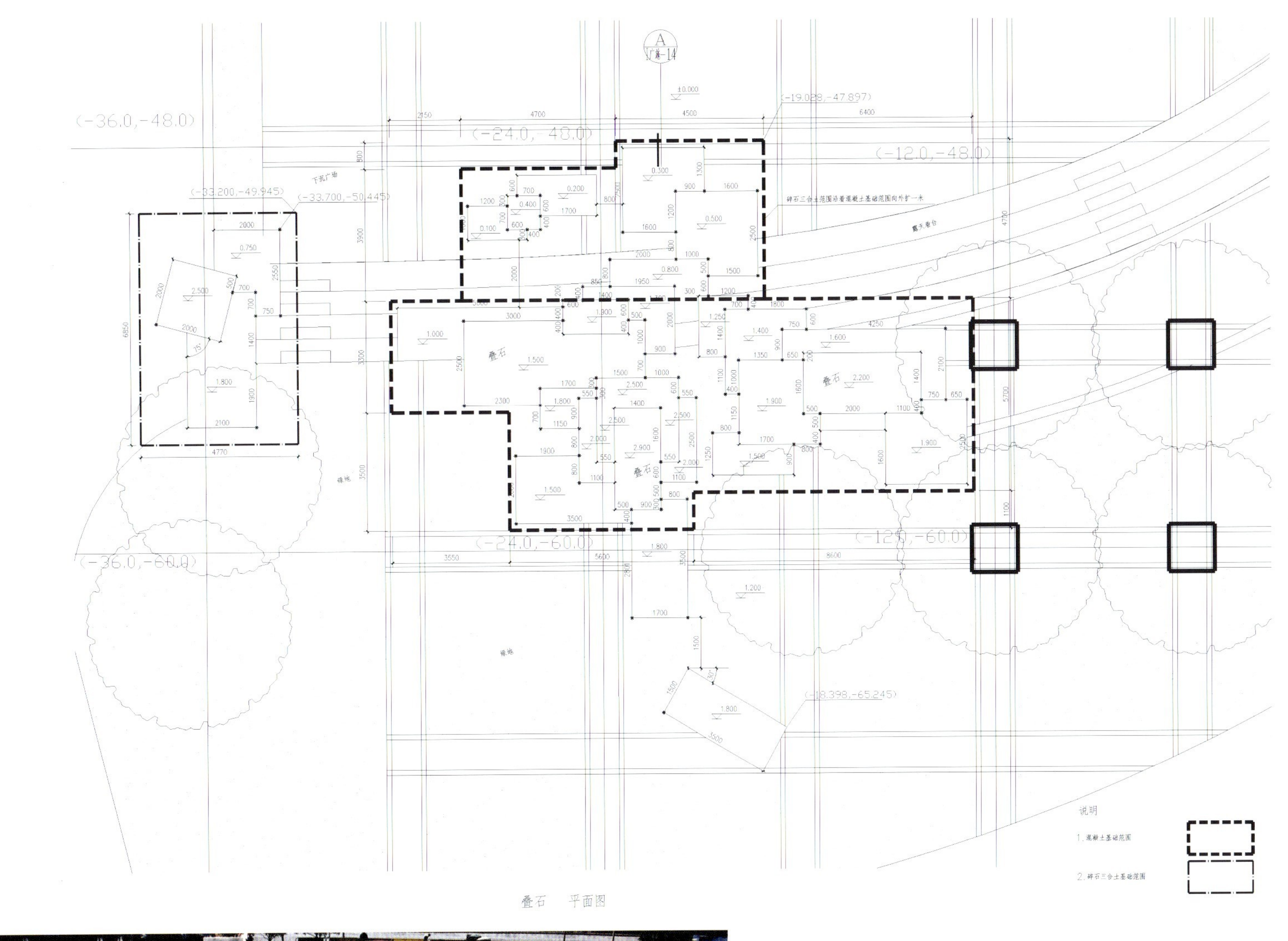


叠石　平面图

露天演艺广场
The amphitheater

露天演艺广场上的观众
The amphitheater occupied

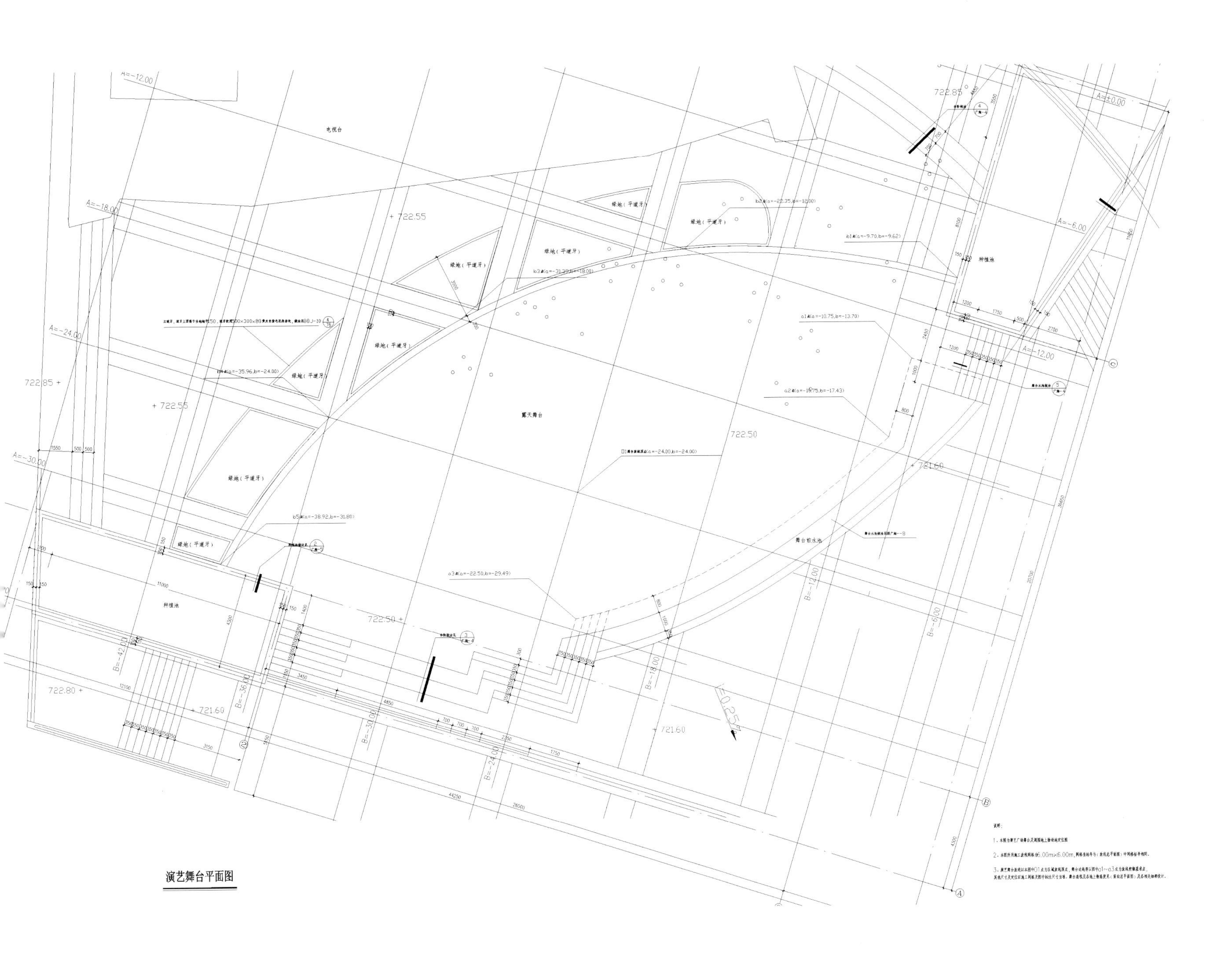
看台
露天舞台
舞台前水池
种植池
绿地（平缓牙）
722.85
722.55
722.50
722.80
721.60
A=±0.00
A=-6.00
A=-12.00
A=-18.00
A=-24.00
A=-30.00
B=-6.00
B=-12.00
B=-18.00
B=-24.00
B=-30.00
B=-36.00
B=-42.00
i=0.25%
说明：
1．本图为演艺广场舞台及周围地上物场地定位图
2．本图所用施工放线网格为6.00m×6.00m，网格坐标号与（放线总平面图）中网格标号相同。
3．演艺舞台放线以本图中O1点为区域放线原点，舞台边缘等以图中a1~a3点为放线控制基准点，
其他尺寸及定位以施工网格及图中标注尺寸为准。舞台高程及各地上物高度见（竖向总平面图）及各相关细部设计。

演艺舞台平面图

花池
台阶
喷泉水池
600x600 浅暖黄色花岗石细磨
浅灰色花岗石粗斩台沿
浅灰色花岗石粗斩台沿
浅灰色花岗石粗磨台
X=2985730.466
Y=134510.962
X=2985726.963
Y=134522.439
R27357
R2000

竹林分隔的后台排演场地
The bamboo encircled practice area

金色天幔(杩槎天幔)：在露天演艺舞台上一侧，是一组金色天幔雕塑，由金属片和青铜柱构成。下垂的天幔源于对阳春三月川西油菜花的体验，而斜立的青铜柱则是源于古老的治水技术杩槎的灵感。这一雕塑成为不同时段、不同人群和不同活动的场景：早晨是跳舞和舞剑妇女们的优选地，晚上则是男人和观光客获得温暖感和激动的地方，那是对阳春三月油菜花体验的回味

Golden Canopy: At the south west corner of the square and on the stage, the golden metal canopy hung on bronze poles was inspired by the rapeseed blossoms and poles in triplets used in irrigation works. It is one of many art works in the square. It becomes an attractive scene for various activities at different time: in the morning, women dance and play *Tai Chi* using it as the theatrical scenes, and in the night, tourists and residents come here to recall the experience of the bright and cheerful atmosphere arisen by the rapeseed blossoms in the warm spring

治水杩槎和川西的油菜花：金色天幔的灵感来源
The poles in triplets used in irrigation works and rapeseed blossoms are the inspiration for the golden canopy

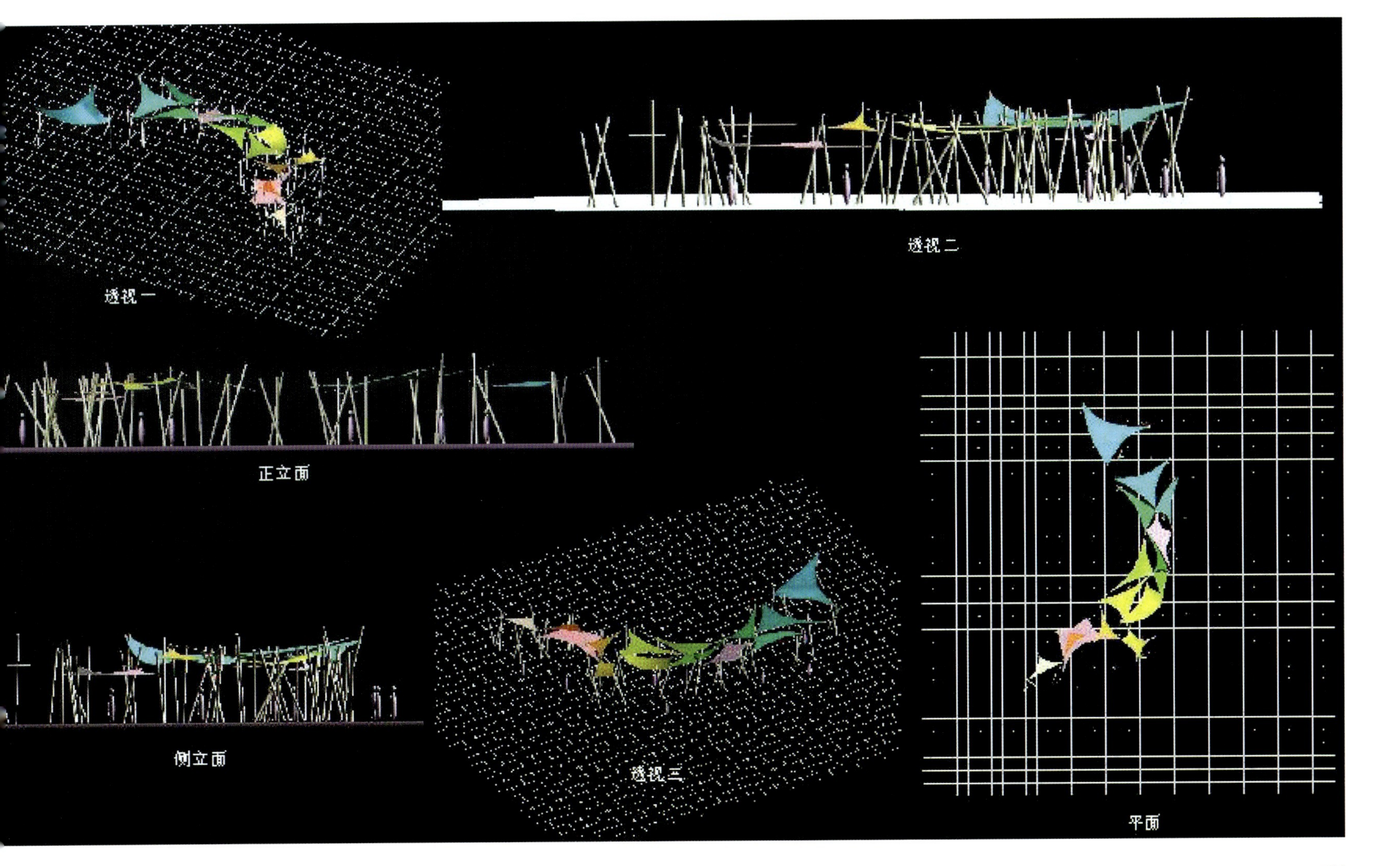

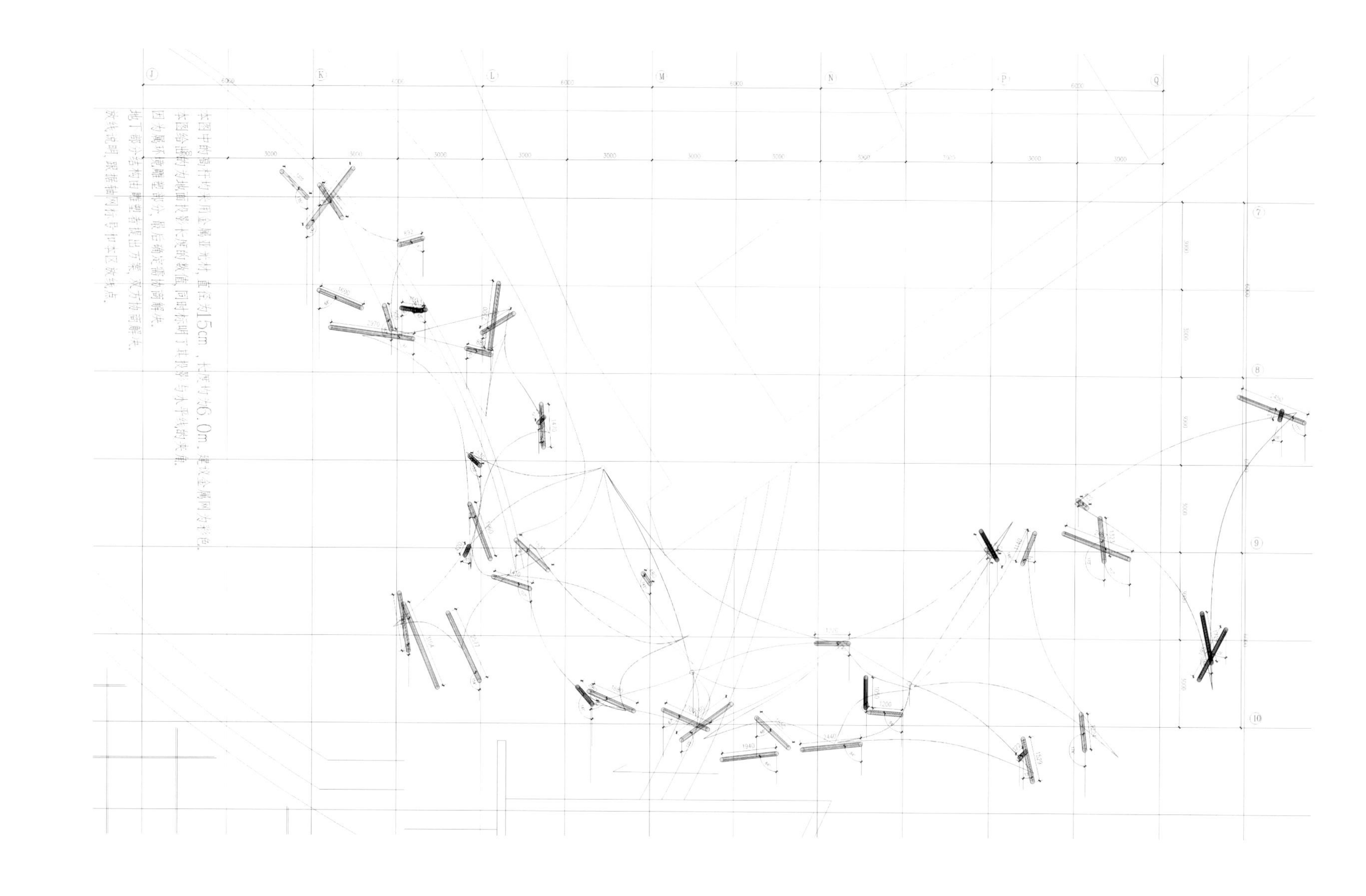


金色天幔的平面布局

The layout of the golden canopy

河边步道：在广场I区之滨水一侧，设计尊重当地人的原有休闲方式，在急流的河边设置大量座凳，一排灯柱与河成一斜角排列。当地的妇女带着日常生活中不可或缺的竹筐来此休闲，不经意间竹筐被融入了预设的、也是源于它们的场景之中。晚上则聚集了吹凉风和听水声的成都人、游客和当地人

River walk: A river walk by one of the canal at the south west side of the square. People site in line along the river as they are used to do, granite lighting columns is in a symbolic form of a bamboo basket. Details of fences also in the similar pattern language, which mingled with the bamboo weaved containers used by the local people. Tourists mingled with local residents sitting and walking along the canal in the night, enjoy the cool winds and the sound of the rapid flows

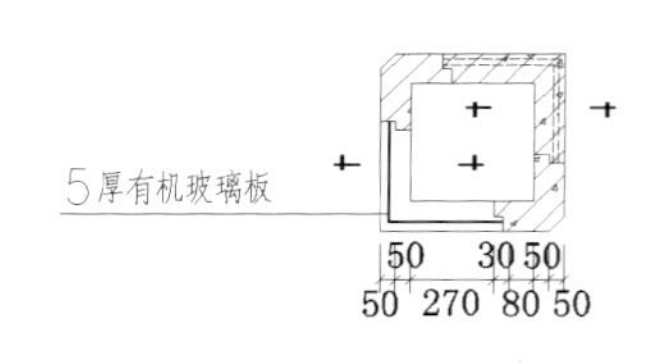


灯柱平面

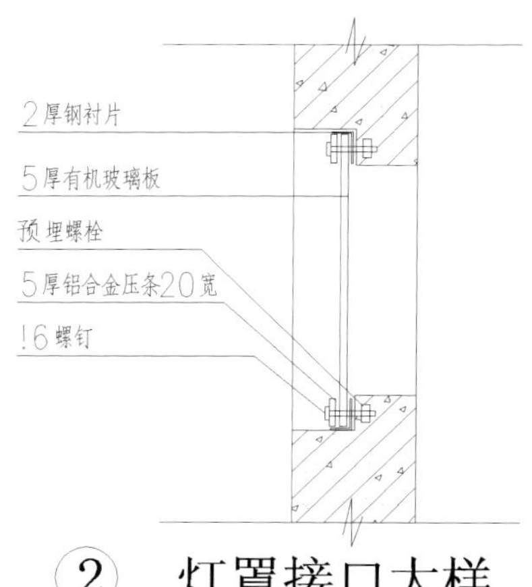


② 灯罩接口大样

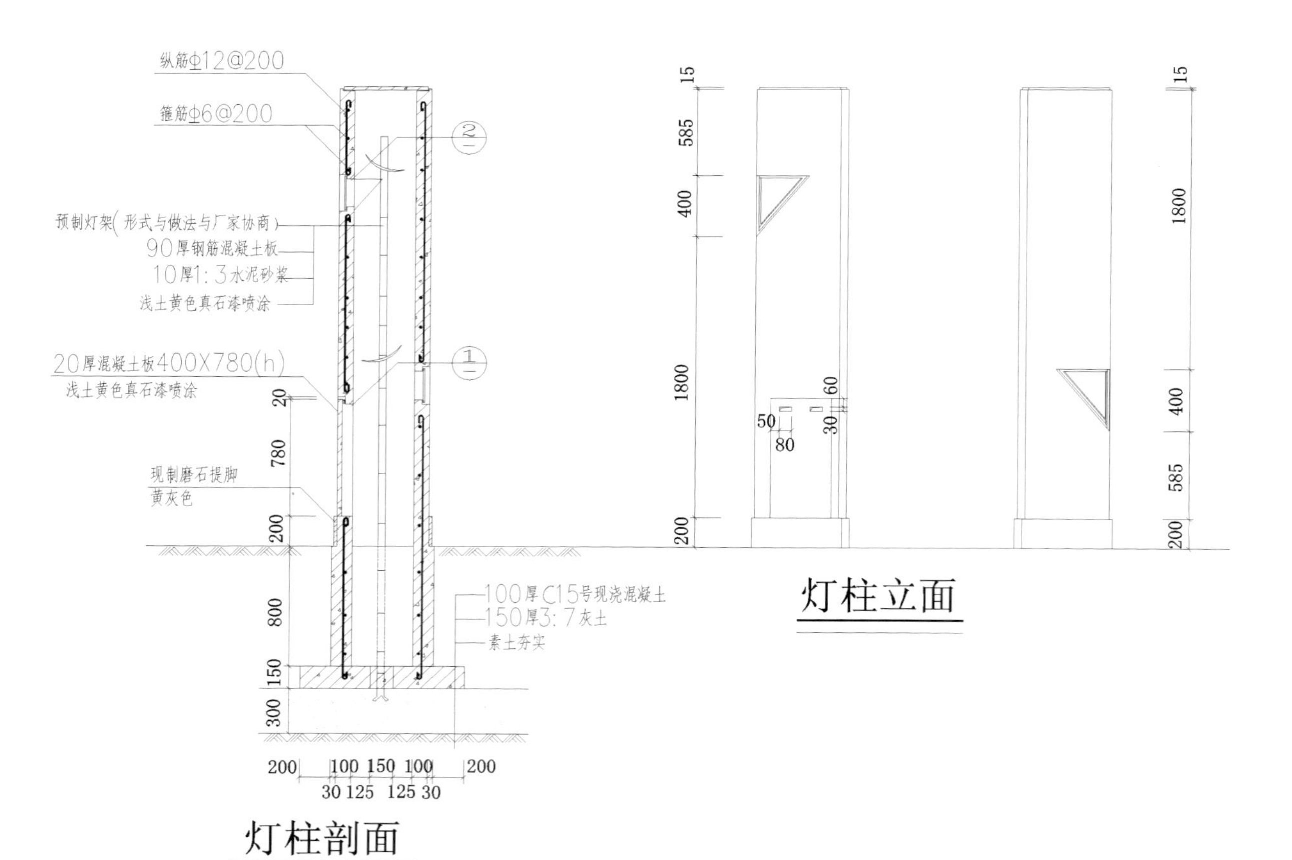


灯柱剖面

灯柱立面

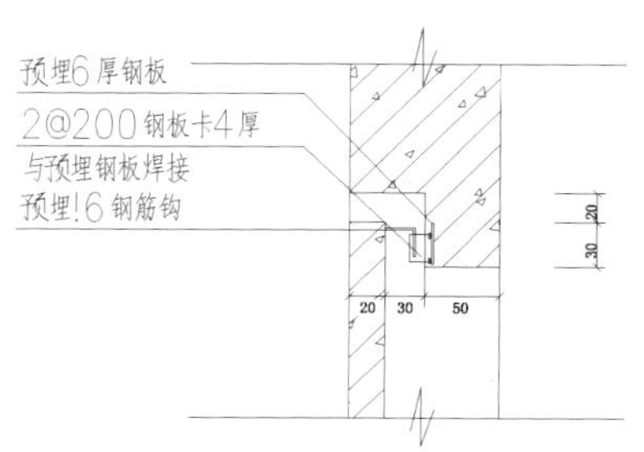


① 活动门接口大样

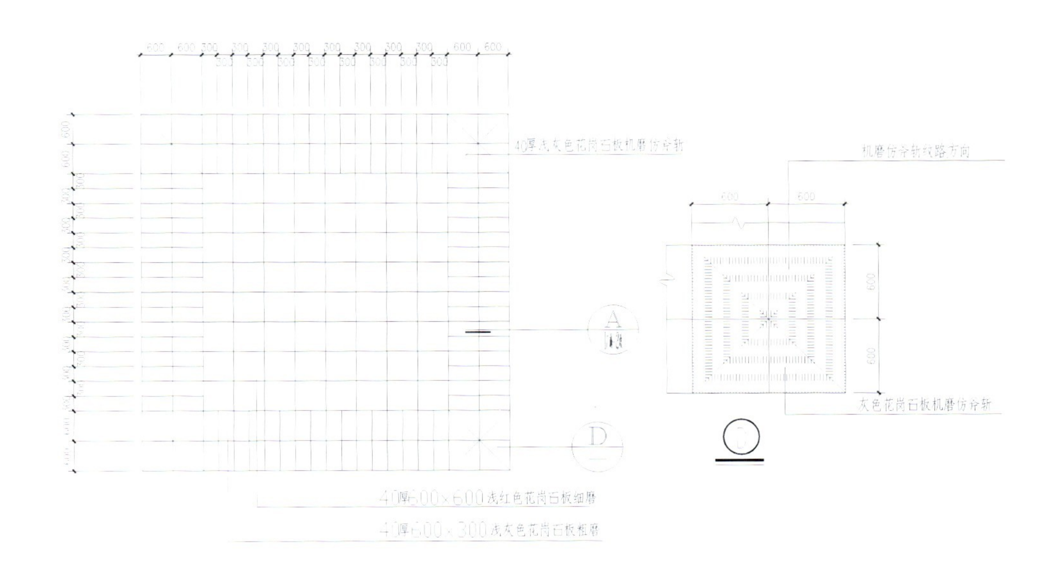


广场地面花岗石铺装平面详图

主题雕塑“投玉入波”：源于竹笼原形，高30m，基底为一波状水景，共同构成“投玉入波”主题。网纹与卵石镂刻相结合，是竹编结构的高度艺术化。沿轴线方向为三个灯柱，与主雕塑同样由花岗石镂刻而成，内有光源。在广场北部有一导水渠，使主题雕塑整体倒影入水。主题雕塑和灯柱立于由卵石和浅水所构成的水景中。

The focal sculpture: 30 meters high, the iconic sculpture and lighting columns on the central axis recall the unique bamboo baskets used in building the Dujiangyan Weir, an icon of the Dujiangyan City. The columns are carved in granite, and lighted inside. The ground is like a riverbed covered with a thin layer of water and pebbles of various sizes, also an inspiration from local river experience, carved in granite. It is a focal point of the whole square. At the north part of the square, the sculpture is reflected in the creek that extends the central axis

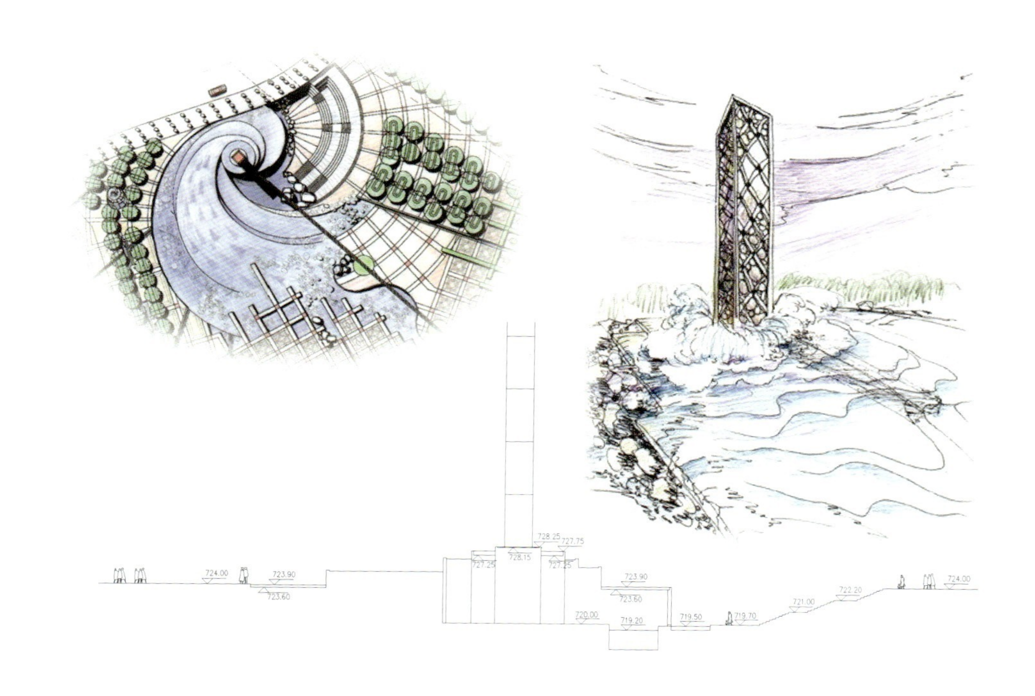

A—A剖面图

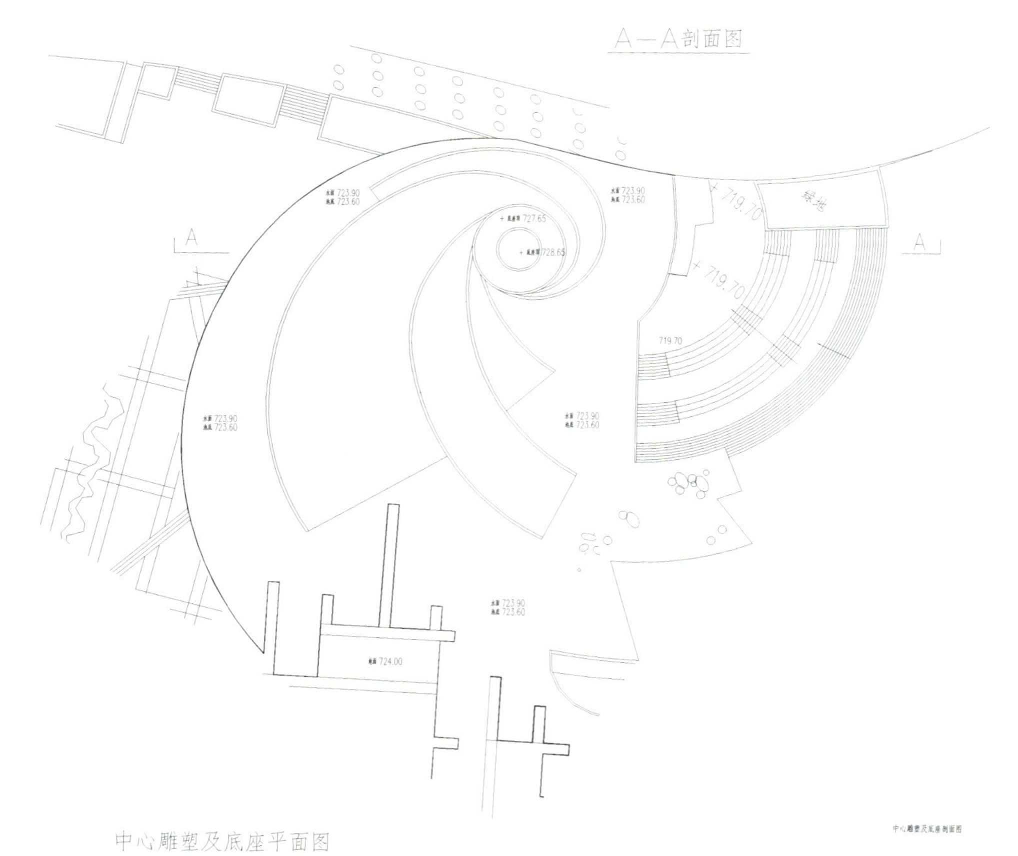

中心雕塑及底座平面图

中心雕塑及底座剖面图

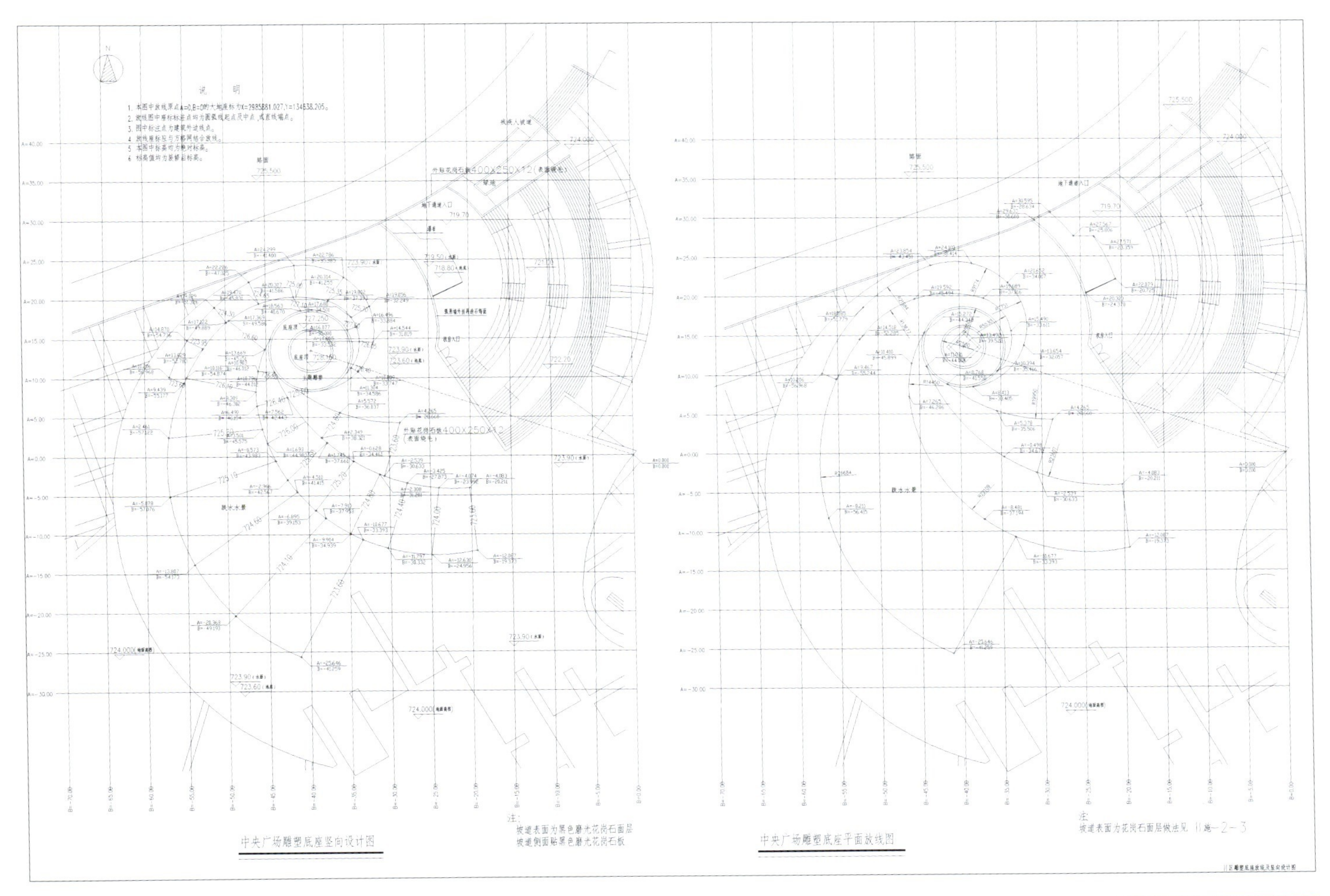
N
说　　明
1. 本图中放线原点A=0,B=0的大地座标为X=2985881.027,Y=134638.205。
2. 放线图中座标标注点均为圆弧线起点及中点，或直线端点。
3. 图中标注点为建筑外缘线点。
4 放线座标应与方格网结合放线。
5 本图中标高均为绝对标高。
6 标高值均为装修后标高。
路面
725.500
中央广场雕塑底座竖向设计图
注：
坡道表面为黑色磨光花岗石面层
坡道侧面贴黑色磨光花岗石板
中央广场雕塑底座平面放线图
注：
坡道表面为花岗石面层做法见 II施－2－3

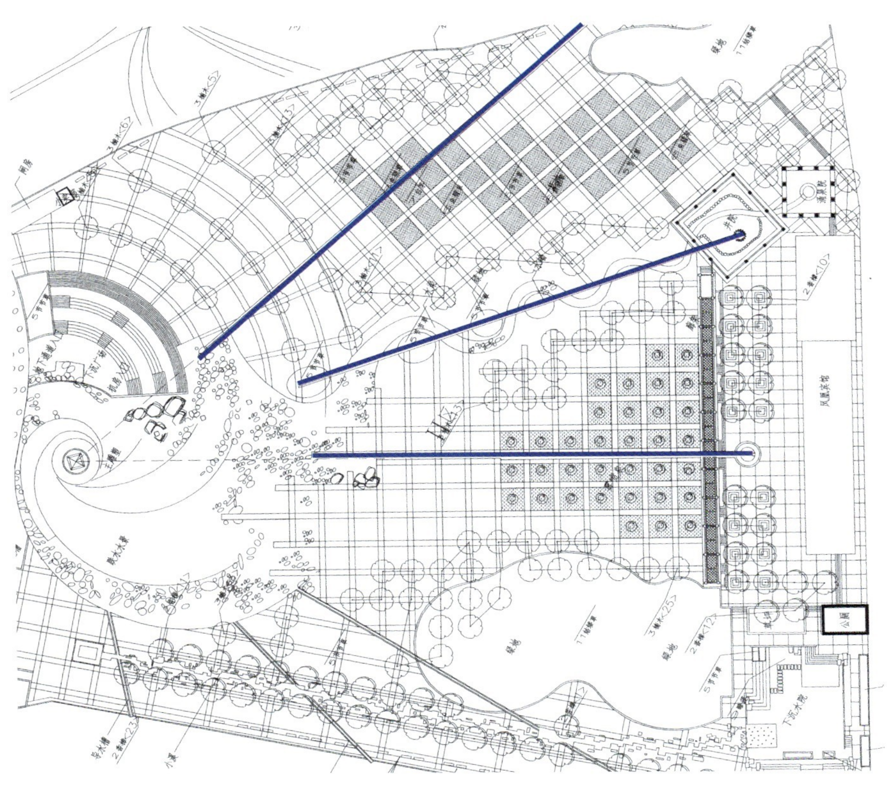

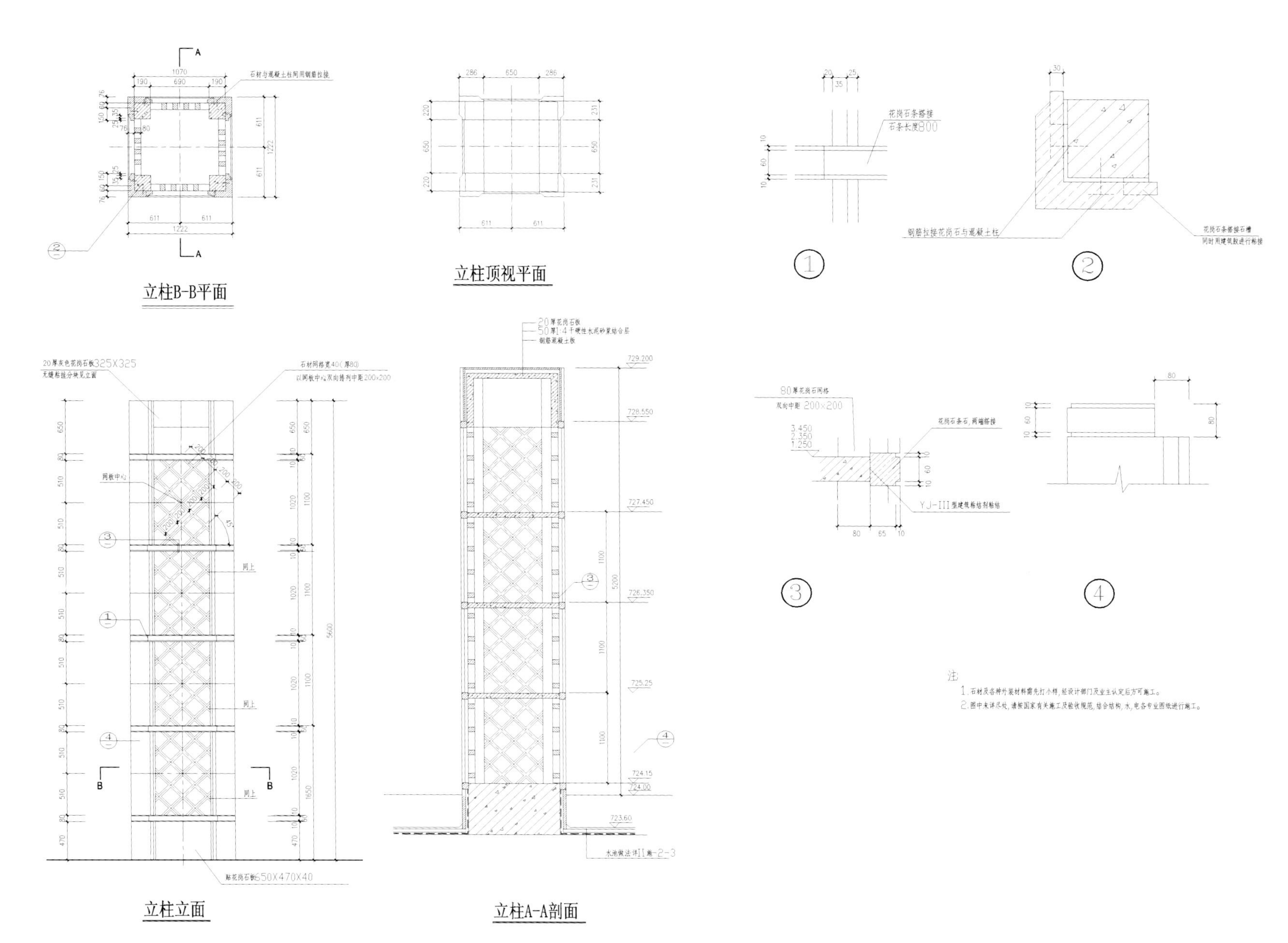
立柱B-B平面
立柱顶视平面
立柱立面
立柱A-A剖面
注
1. 石材及各种外装材料需先打小样，经设计部门及业主认定后方可施工。
2. 图中未详尽处，请按国家有关施工及验收规范，结合结构，水，电各专业图纸进行施工。

网纹水面：在主题雕塑下为多条水道呈水涡状，基地微起众多鱼嘴，水流经过，泛起无数网纹，编制出一个极富动感的水纹“竹笼。”细微处在暗示一个久远亲切的故事

A weaved water surface: A detail of the water feature made of many fish mouths that multiply as water flows down the whirl shaped water way at the base of the focal sculpture. The waves created by the small fish mouths weave a delicate and dynamic water surface that reflects the light. This was inspired from the irrigation works and the bamboo weaving technique used in the local handcrafts for daily uses

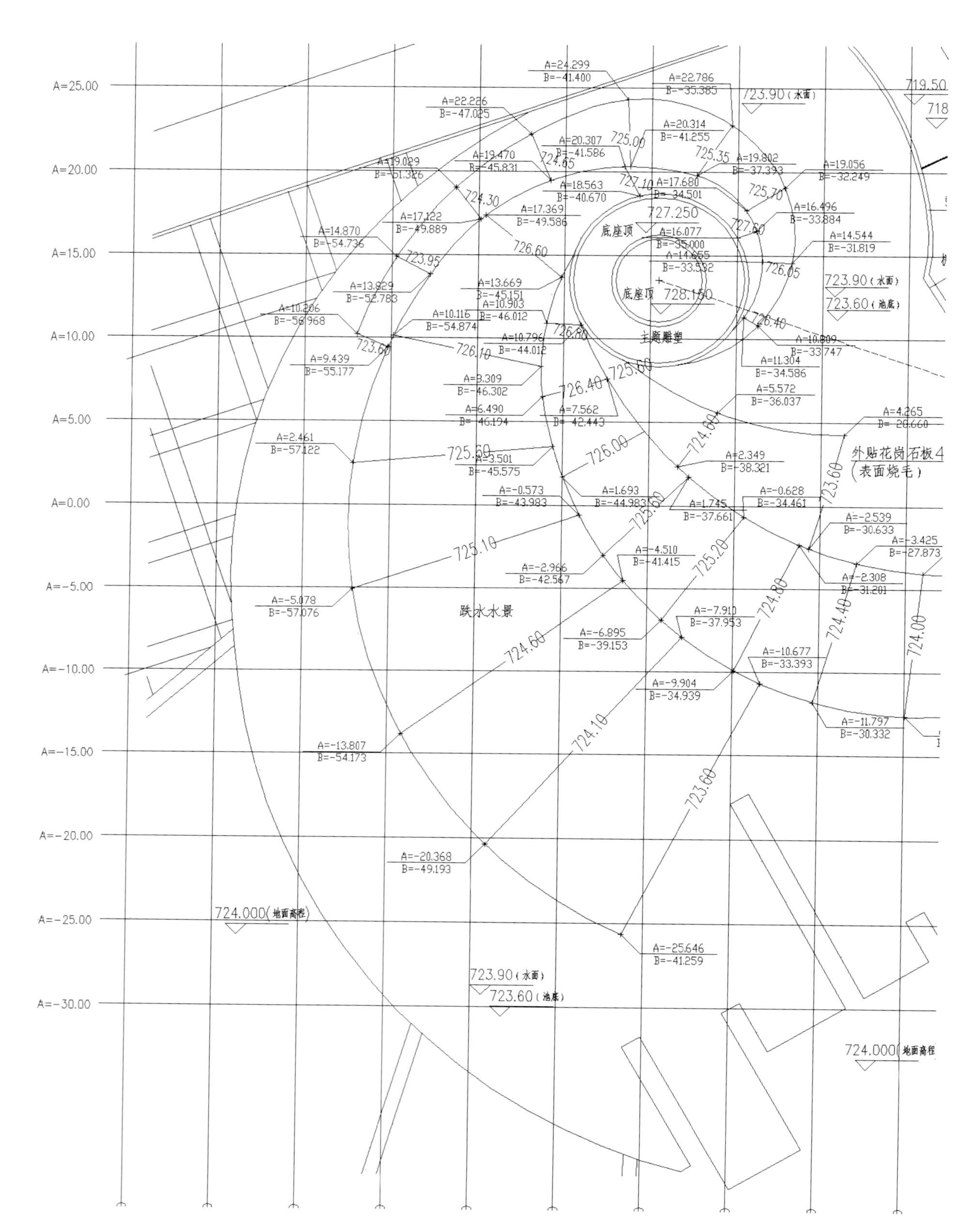

网纹水面设计图

The grading of the weaved water surface

中心轴线：中部的对角向轴线由主雕、漏墙、三个灯柱和一条蜿蜒盘曲的溪流构成。漏墙实际上是一条导水槽，长约100m，同时起到分割广场空间的作用。自南端沿漏墙北望，一注银水自天而来，落入井院之中。川西乡土景观中的导水槽、建筑格栅、民居的井院格局和体验，和“逢正抽心，遇湾截角”的治水技术，都给设计带来了灵感

The central axis: view from the center southwards, made of a carved stone wall, three light columns, and a meandering creek in red sand rock. The red rock creek is typically seen in the surrounding mountains. On the top of the wall is a flume that can be seen at south end as if water comes far away from the Weir, implies the Chinese poem “stream pour from the sky”. Water falls into a courtyard encircled with stone columns and disappear into a well. The vernacular landscape elements and experience such as aqueduct, window grid, courtyard are all sources of design inspiration

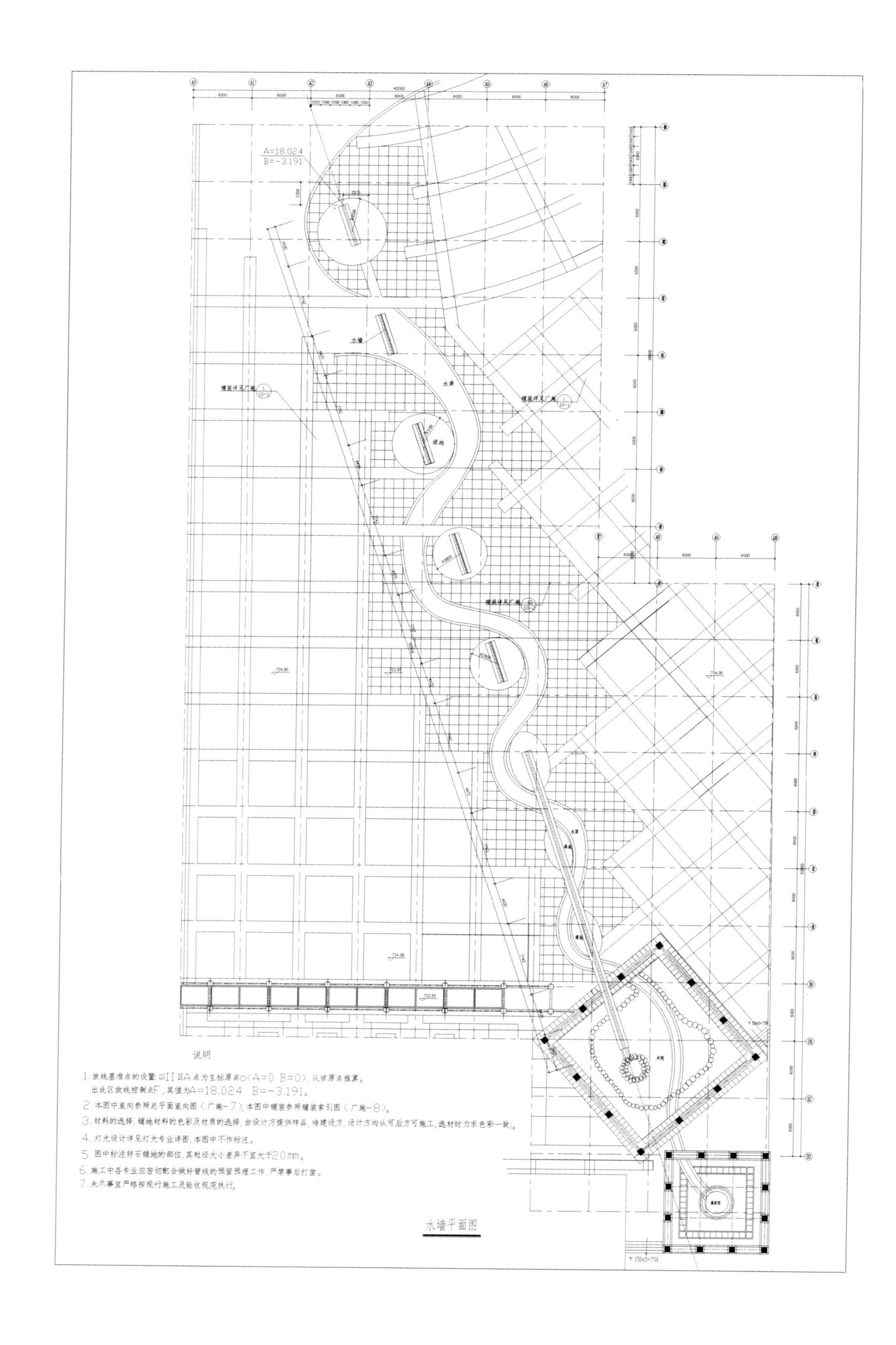

A=18.024
B=-3.191
水景
绿地
说明
1.放线基准点的设置:以II区A点为坐标原点o(A=0 B=0),从该原点推算。
出此区放线控制点F,其值为A=18.024 B=-3.191。
2.本图中竖向参照总平面竖向图(广施-7),本图中铺装参照铺装索引图(广施-8)。
3.材料的选择,铺地材料的色彩及材质的选择,由设计方提供样品,待建设方,设计方均认可后方可施工,选材时力求色彩一致;。
4.灯光设计详见灯光专业详图,本图中不作标注。
5.图中标注卵石铺地的部位,其粒径大小差异不宜大于20mm。
6.施工中各专业应密切配合做好管线的预留预埋工作,严禁事后打凿。
7.未尽事宜严格按现行施工及验收规范执行。
水墙平面图

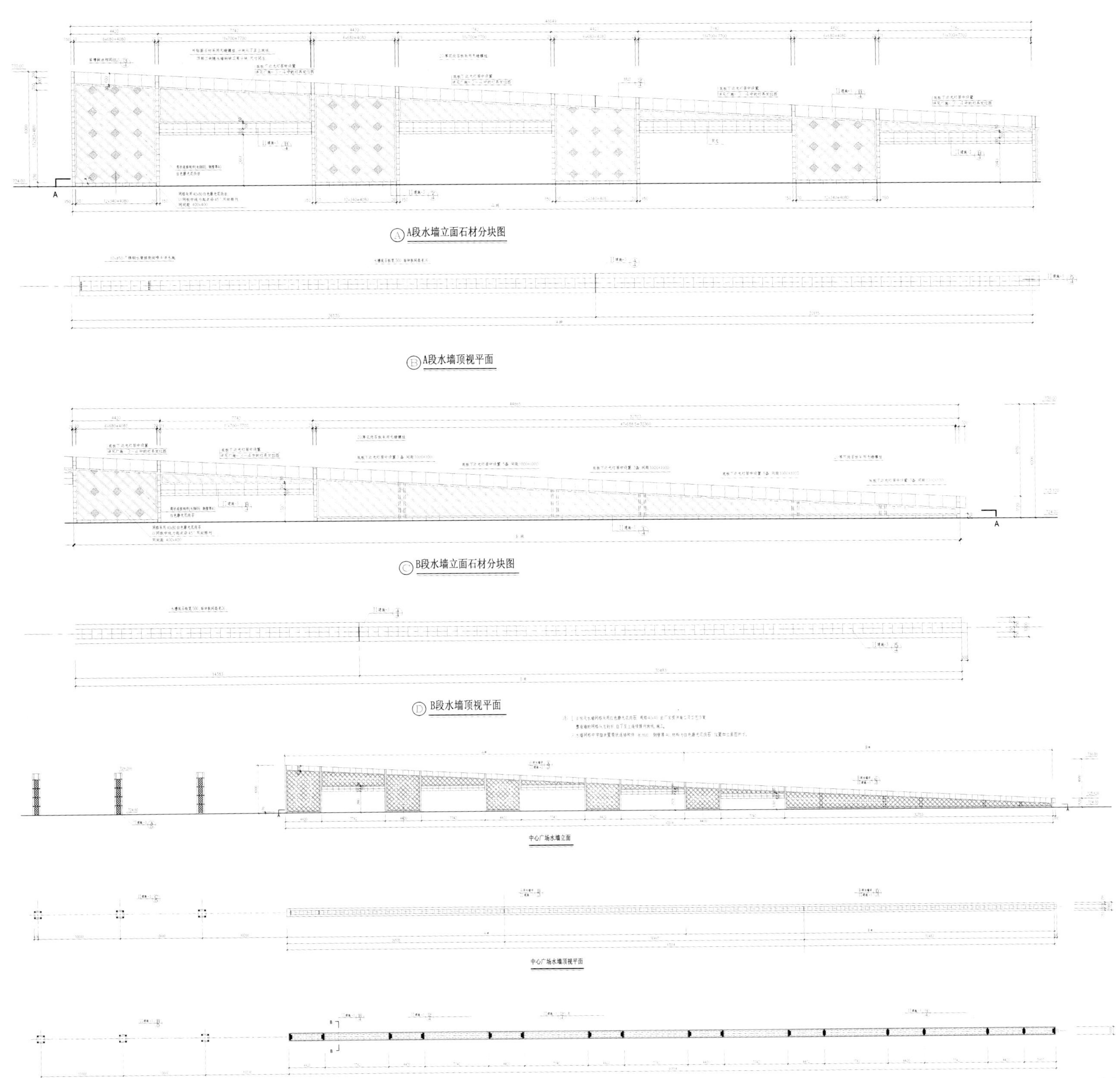
Ⓐ A段水墙立面石材分块图
Ⓑ A段水墙顶视平面
Ⓒ B段水墙立面石材分块图
Ⓓ B段水墙顶视平面
中心广场水墙立面
中心广场水墙顶视平面
中心广场水墙A-A平面

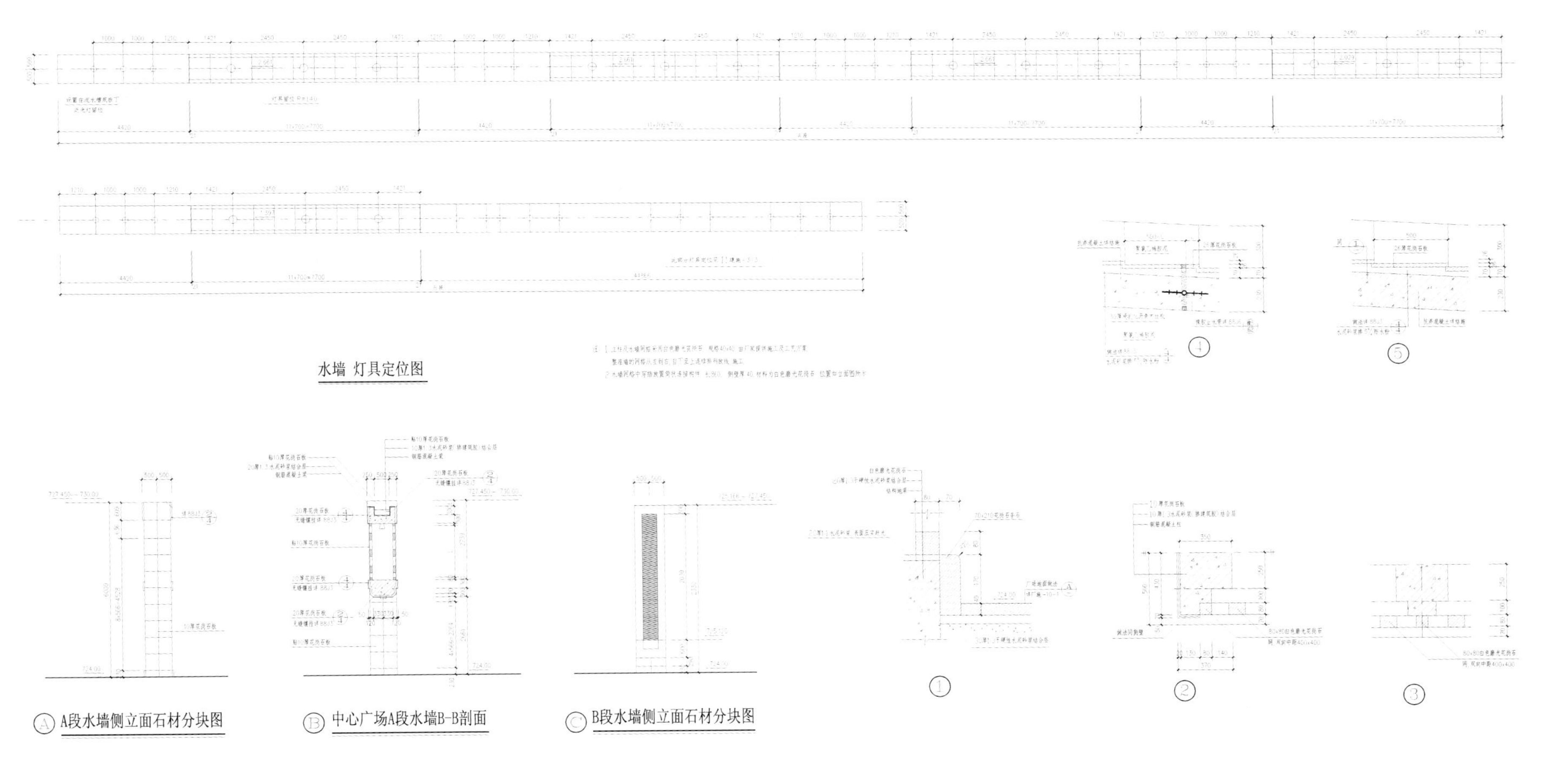

水墙 灯具定位图
A A段水墙侧立面石材分块图
B 中心广场A段水墙B-B剖面
C B段水墙侧立面石材分块图

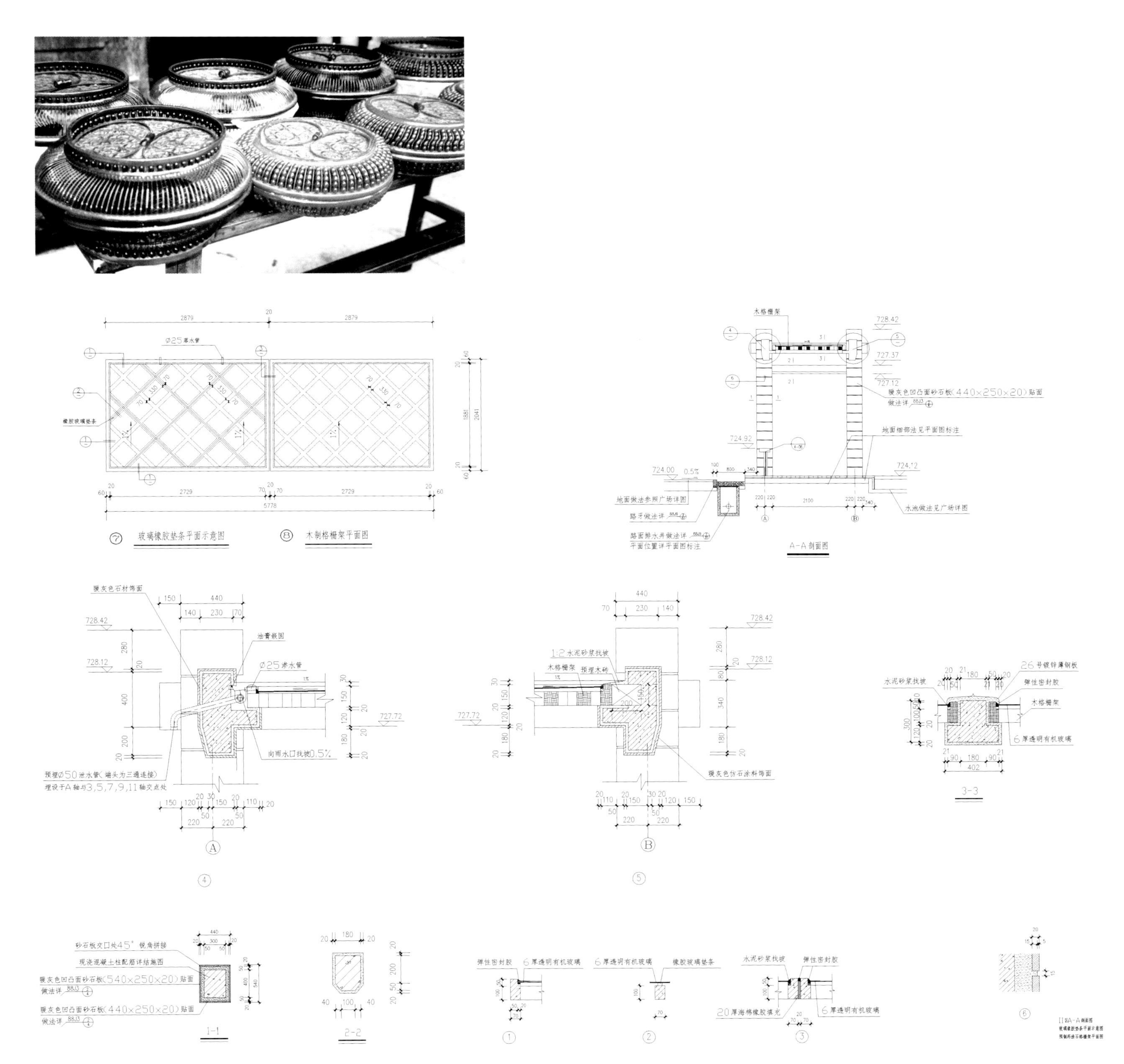

Ø25 滤水管
橡胶玻璃垫条
2879
20
2729
5778
⑦ 玻璃橡胶垫条平面示意图
⑧ 木制格栅架平面图
木格栅架
728.42
727.37
727.12
腰灰色凹凸面砂石板(440×250×20)贴面
地面细部法见平面图标注
724.92
724.00
0.5%
724.12
地面做法参照广场详图
路牙做法详
路面排水井做法详
平面位置详平面图标注
水池做法见广场详图
A-A 剖面图
腰灰色石材饰面
油膏嵌固
Ø25 滤水管
向雨水口找坡0.5%
预埋Ø50泄水管(端头为三通连接)
埋设于A轴与3,5,7,9,11轴交点处
727.72
728.12
1:2水泥砂浆找坡
木格栅架
预埋木砖
腰灰色仿石涂料饰面
26号镀锌薄钢板
水泥砂浆找坡
弹性密封胶
木格栅架
6厚透明有机玻璃
402
3-3
砂石板交口处45°锐角拼接
现浇混凝土柱配筋详结施图
腰灰色凹凸面砂石板(540×250×20)贴面
腰灰色凹凸面砂石板(440×250×20)贴面
1-1
2-2
弹性密封胶
6厚透明有机玻璃
橡胶玻璃垫条
水泥砂浆找坡
20厚海棉橡胶填充

啤酒节

广场上的漏墙起到空间分隔的作用，使开敞的广场变得更加丰富
The carved stone wall acts as a space defining screen wall that enriches the experience of the otherwise monotonous open space

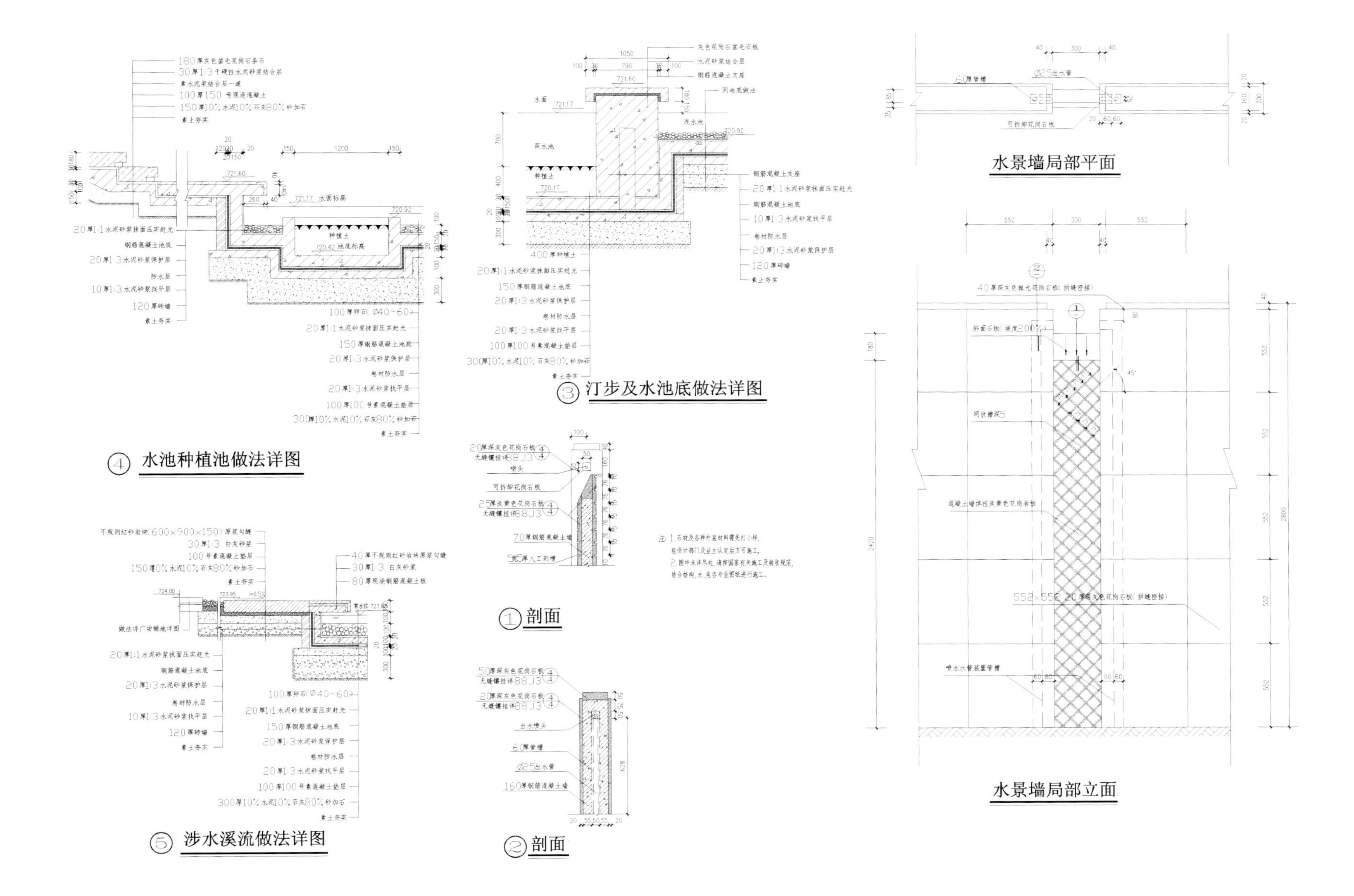

④ 水池种植池做法详图
③ 汀步及水池底做法详图
⑤ 涉水溪流做法详图
① 剖面
② 剖面
水景墙局部平面
水景墙局部立面

亲水性和玩性设计：纵穿场地的四条河流水势湍急，可及性差。本设计将河中之水提上广场，以供人们嬉戏。中央浅水池中的红砂岩卵石取之当地河流，浅池设计本身也从当地溪流的卵石滩景观中获得灵感

Making water accessible and playful: The slope of the four irrigation canals are steep with rapid water flows, and are not allowed to be modified due to water management regulation. The design made use of the elevation difference to divert water using traditional ways from the upper reach of the canals and create a creek scattered with red stone found in the local river, allowing the water to become accessible, touchable and playful

亲水体验：从川西乡土景观的石埠和亲水生活中获得灵感

The experience of water: inspired by the vernacular landscape of rural Sichuan

Teen

场院体验：从川西民居的场院中获得灵感
The experience of courtyard: inspired by the vernacular landscape of Sichuan

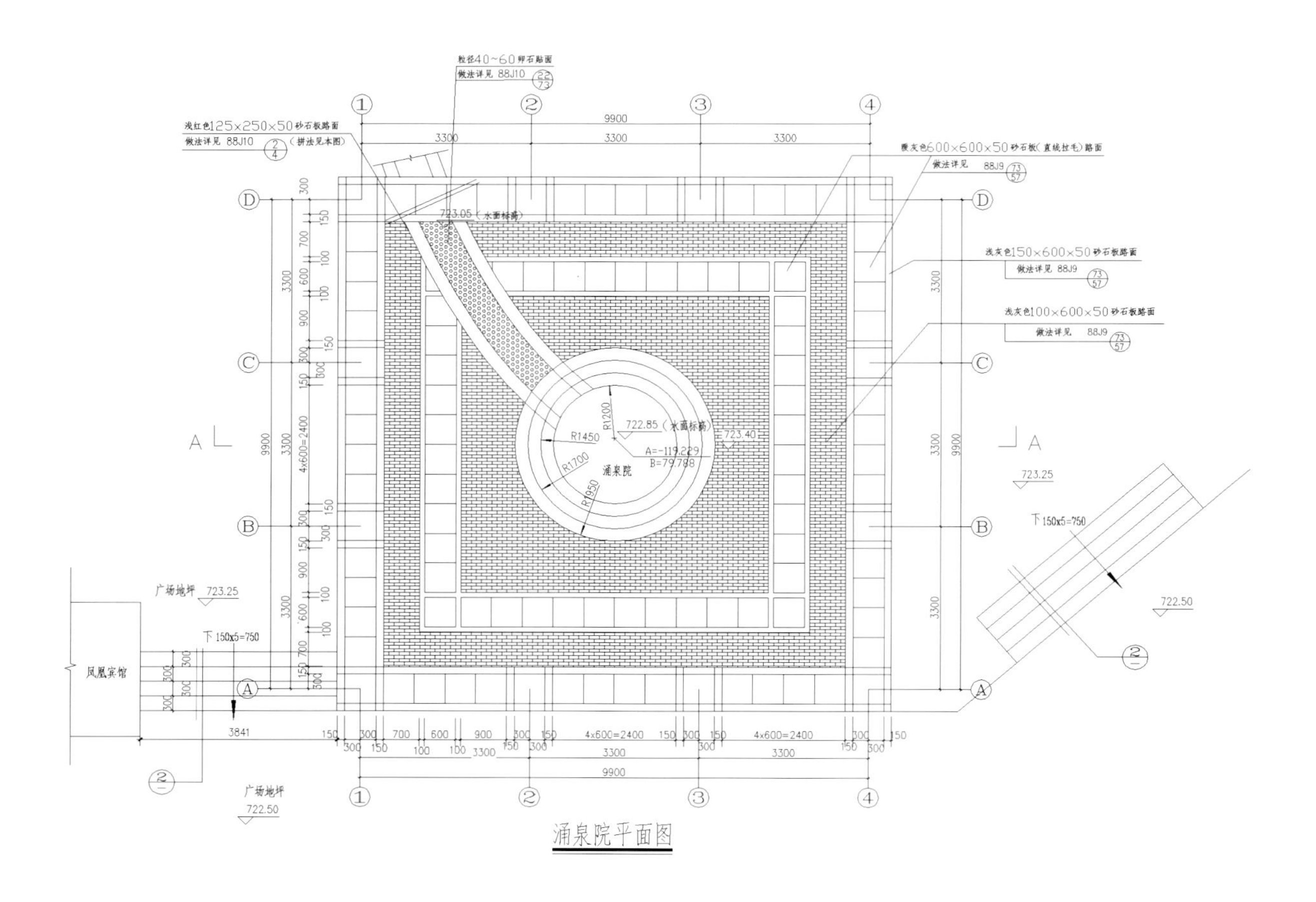
涌泉院平面图
风凰宾馆
广场地坪
723.25
广场地坪
722.50
涌泉院
9900
3300
3300
3300

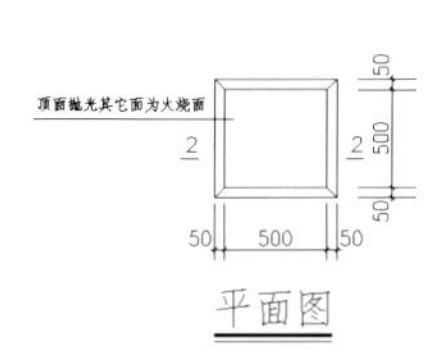
平面图

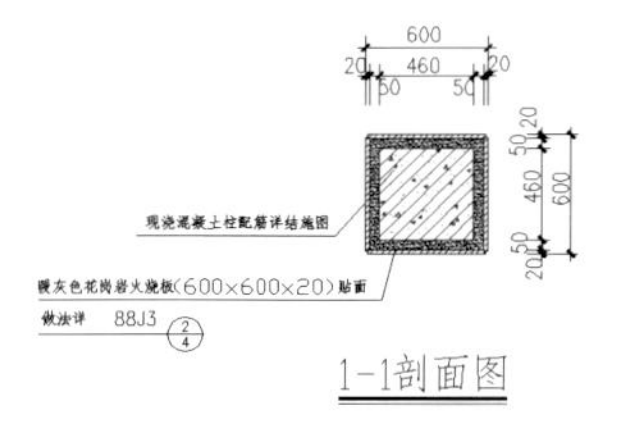
1-1剖面图

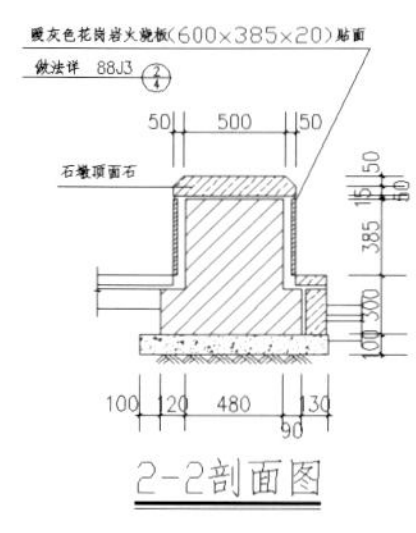
2-2剖面图

顶面石

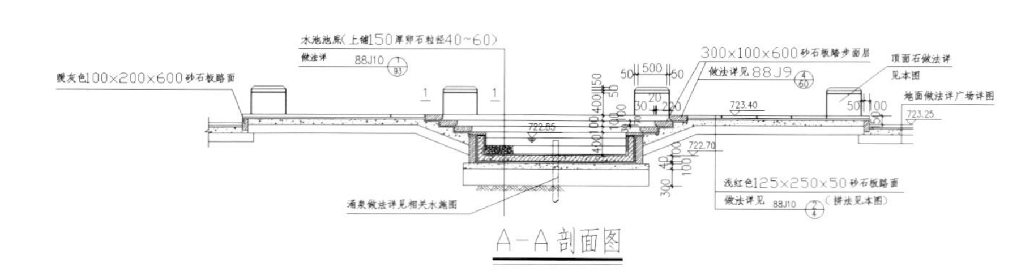
A-A剖面图

下沉式水景广场：作为II区广场西侧溪流的汇，提供一个安静而富有情趣的休息空间
The sunken water garden: it collects water from the stream at the west side of the square

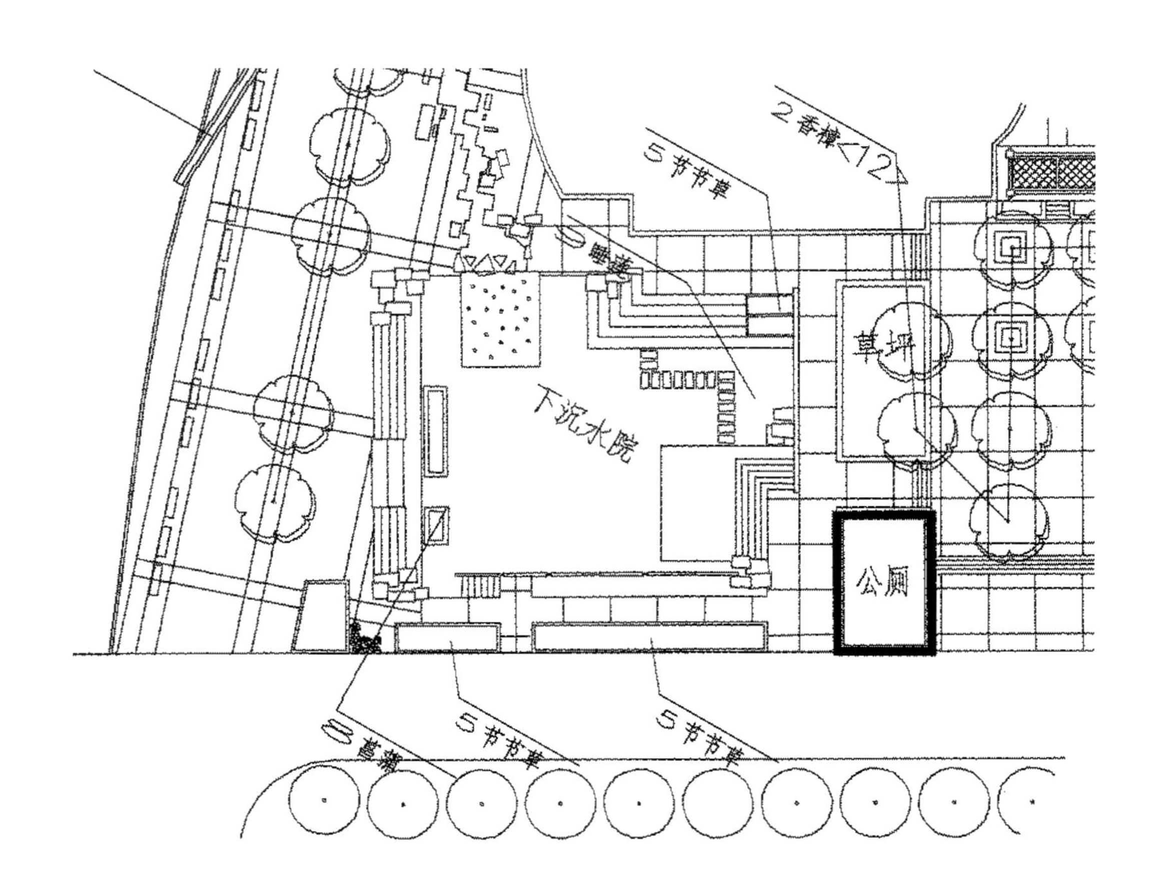
5节节草
2香樟<12>
睡莲
下沉水院
草坪
公厕
5节节草
5节节草

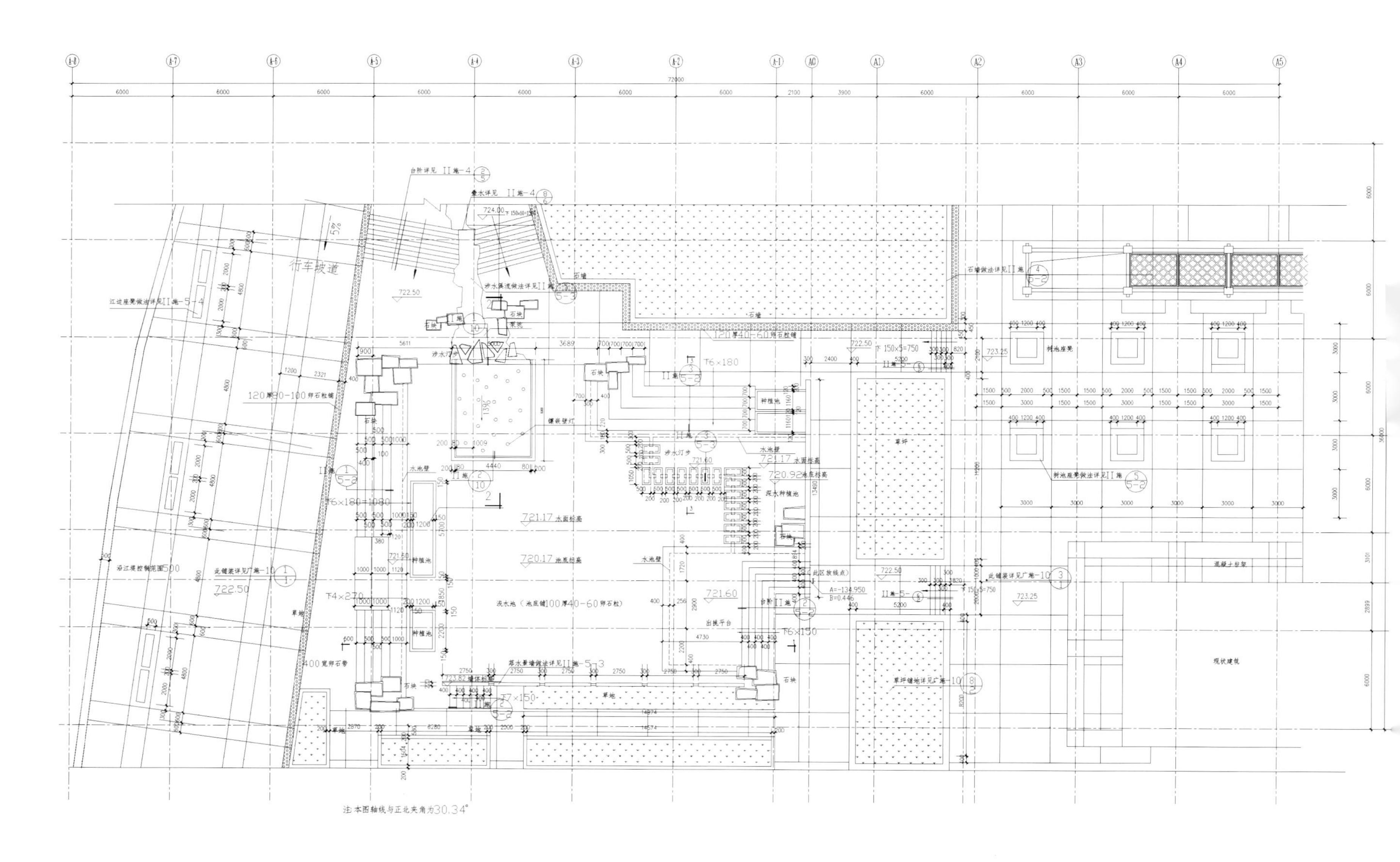


下沉水池平面详图

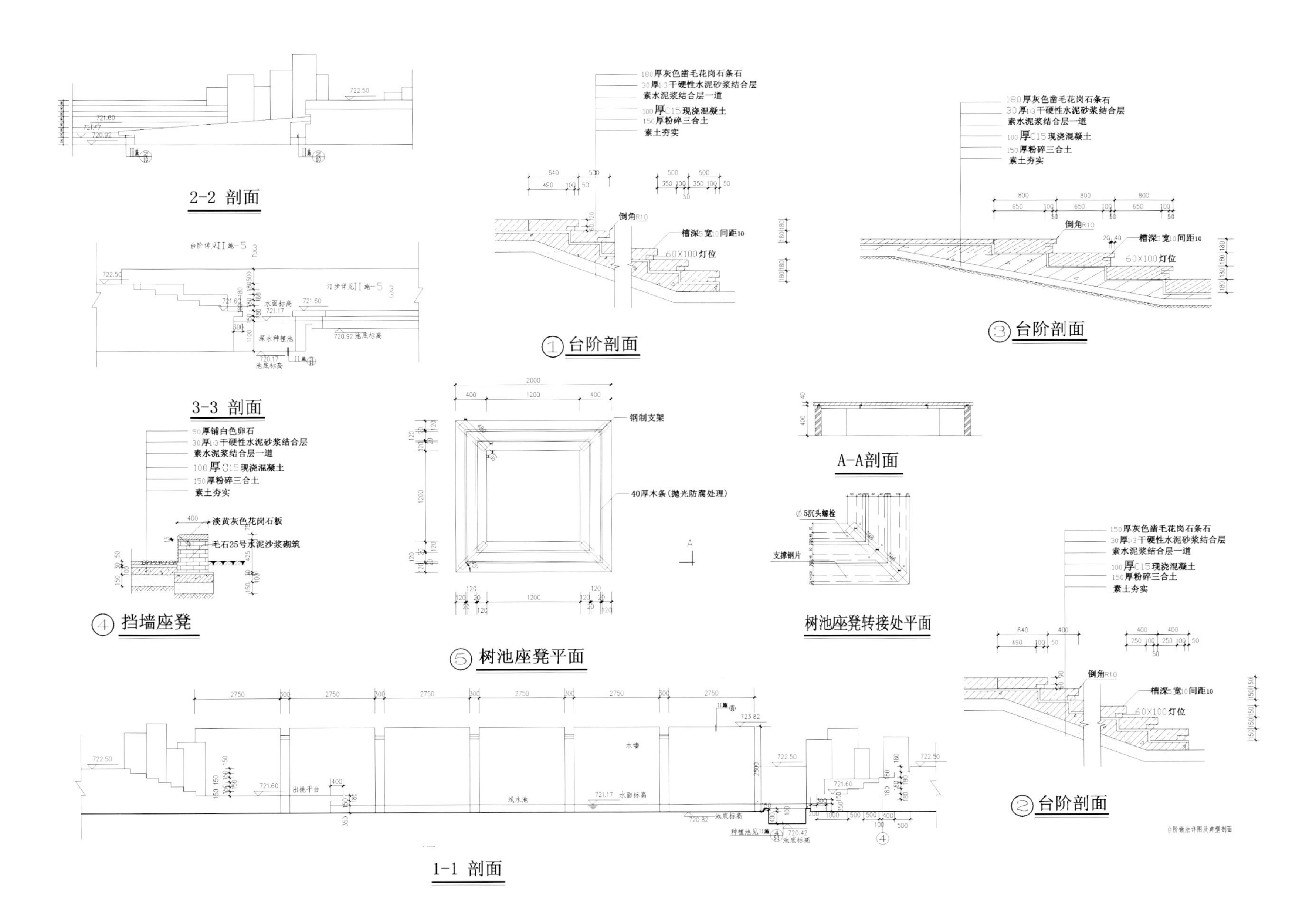
2-2 剖面
180厚灰色凿毛花岗石条石
30厚1:3干硬性水泥砂浆结合层
素水泥浆结合层一道
100厚C15现浇混凝土
150厚粉碎三合土
素土夯实
倒角R10
槽深5宽10间距10
60X100灯位
①台阶剖面
③台阶剖面
3-3 剖面
50厚铺白色卵石
淡黄灰色花岗石板
毛石25号水泥沙浆砌筑
④挡墙座凳
钢制支架
40厚木条（抛光防腐处理）
⑤树池座凳平面
A-A剖面
∅5沉头螺栓
支撑钢片
树池座凳转接处平面
150厚灰色凿毛花岗石条石
②台阶剖面
1-1 剖面

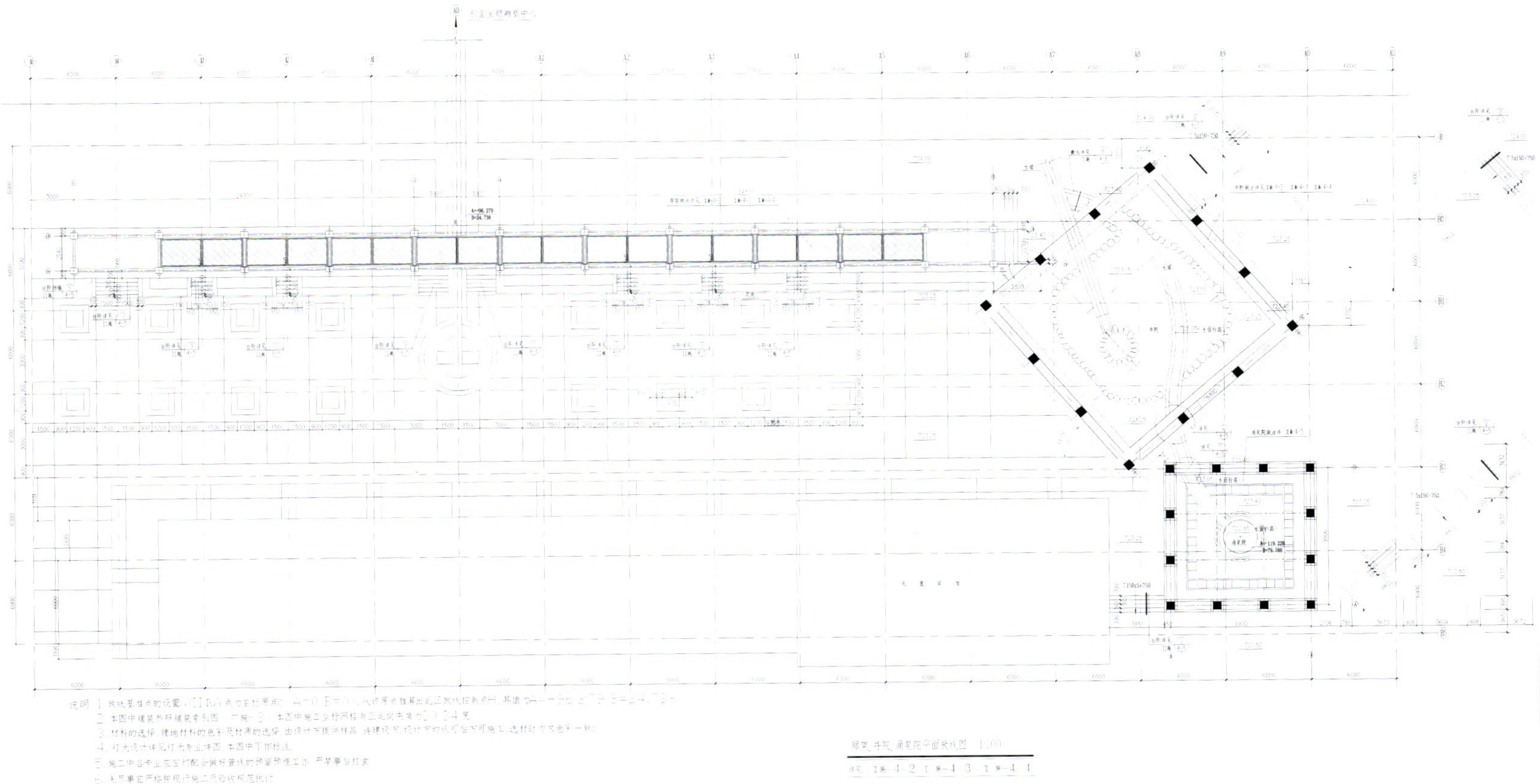

广场南端长廊分割出动与静的两个空间，长廊以北是热闹的水景和活动中心，长廊以南则是安静的林下休闲场所。南侧的带状下沉广场与主广场有1m左右的高差，密植樟树，大量座凳与树池结合，构成一个安静的休憩空间

The corridor at the south end of the square defines two spaces: the active space at the north and the passive at the south, the elevation of which is about 1 meter lower than the north，thus creates a quiet sunken space covered with camphor trees

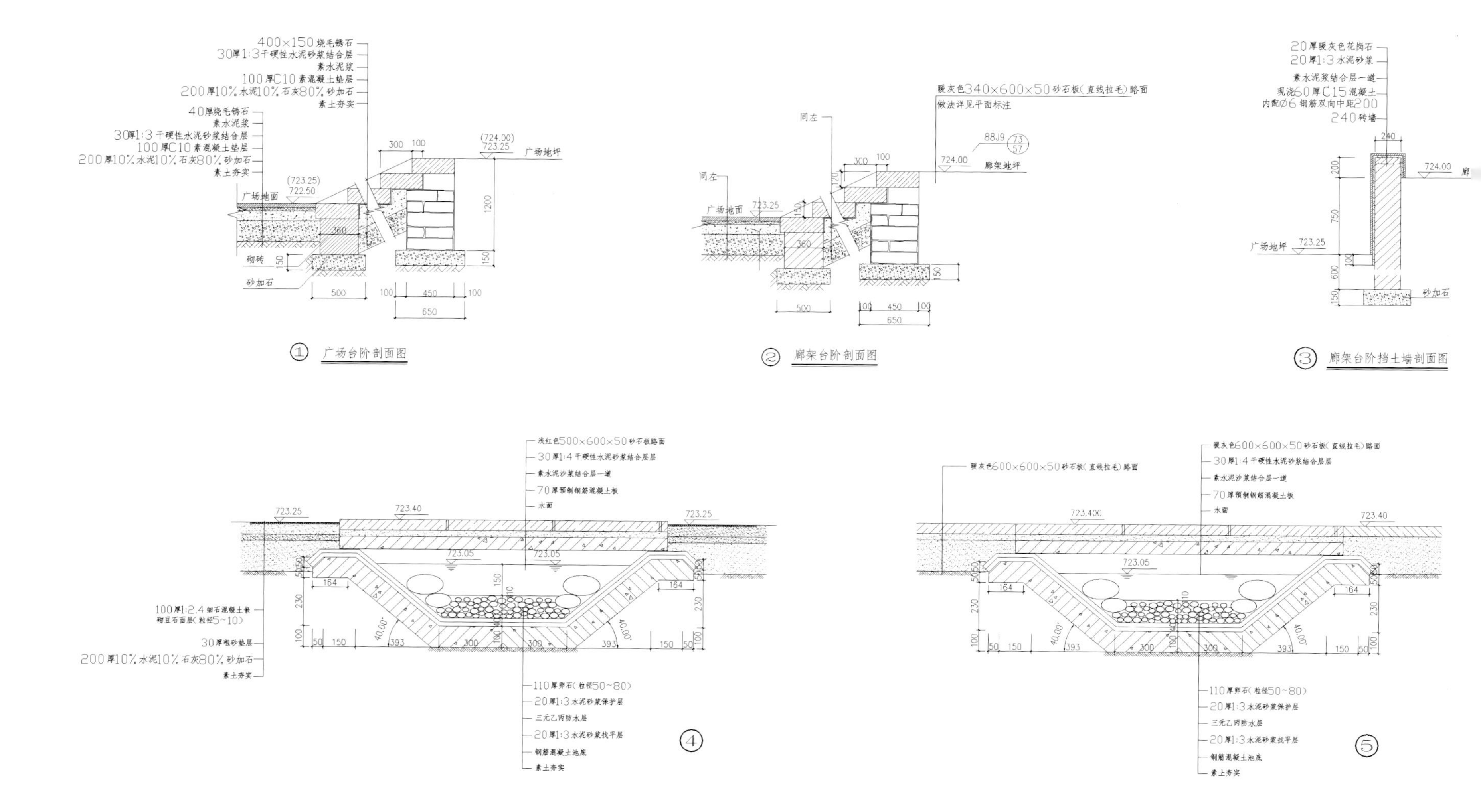

400×150烧毛锈石
30厚1:3干硬性水泥砂浆结合层
素水泥浆
100厚C10素混凝土垫层
200厚10%水泥10%石灰80%砂加石
素土夯实
40厚烧毛锈石
素水泥浆
30厚1:3干硬性水泥砂浆结合层
100厚C10素混凝土垫层
200厚10%水泥10%石灰80%砂加石
素土夯实
广场地面
广场地坪
砌砖
砂加石
① 广场台阶剖面图
同左
同左
广场地面
暖灰色340×600×50砂石板(直线拉毛)路面
做法详见平面标注
88J9
廊架地坪
② 廊架台阶剖面图
20厚暖灰色花岗石
20厚1:3水泥砂浆
素水泥浆结合层一道
现浇60厚C15混凝土
内配Ø6钢筋双向中距200
240砖墙
广场地坪
砂加石
③ 廊架台阶挡土墙剖面图
浅红色500×600×50砂石板路面
30厚1:4干硬性水泥砂浆结合层层
素水泥沙浆结合层一道
70厚预制钢筋混凝土板
水面
100厚1:2.4细石混凝土嵌
铺豆石面层(粒径5~10)
30厚粗砂垫层
200厚10%水泥10%石灰80%砂加石
素土夯实
110厚卵石(粒径50~80)
20厚1:3水泥砂浆保护层
三元乙丙防水层
20厚1:3水泥砂浆找平层
钢筋混凝土池底
素土夯实
④
暖灰色600×600×50砂石板(直线拉毛)路面
暖灰色600×600×50砂石板(直线拉毛)路面
30厚1:4干硬性水泥砂浆结合层层
素水泥沙浆结合层一道
70厚预制钢筋混凝土板
水面
110厚卵石(粒径50~80)
20厚1:3水泥砂浆保护层
三元乙丙防水层
20厚1:3水泥砂浆找平层
钢筋混凝土池底
素土夯实
⑤

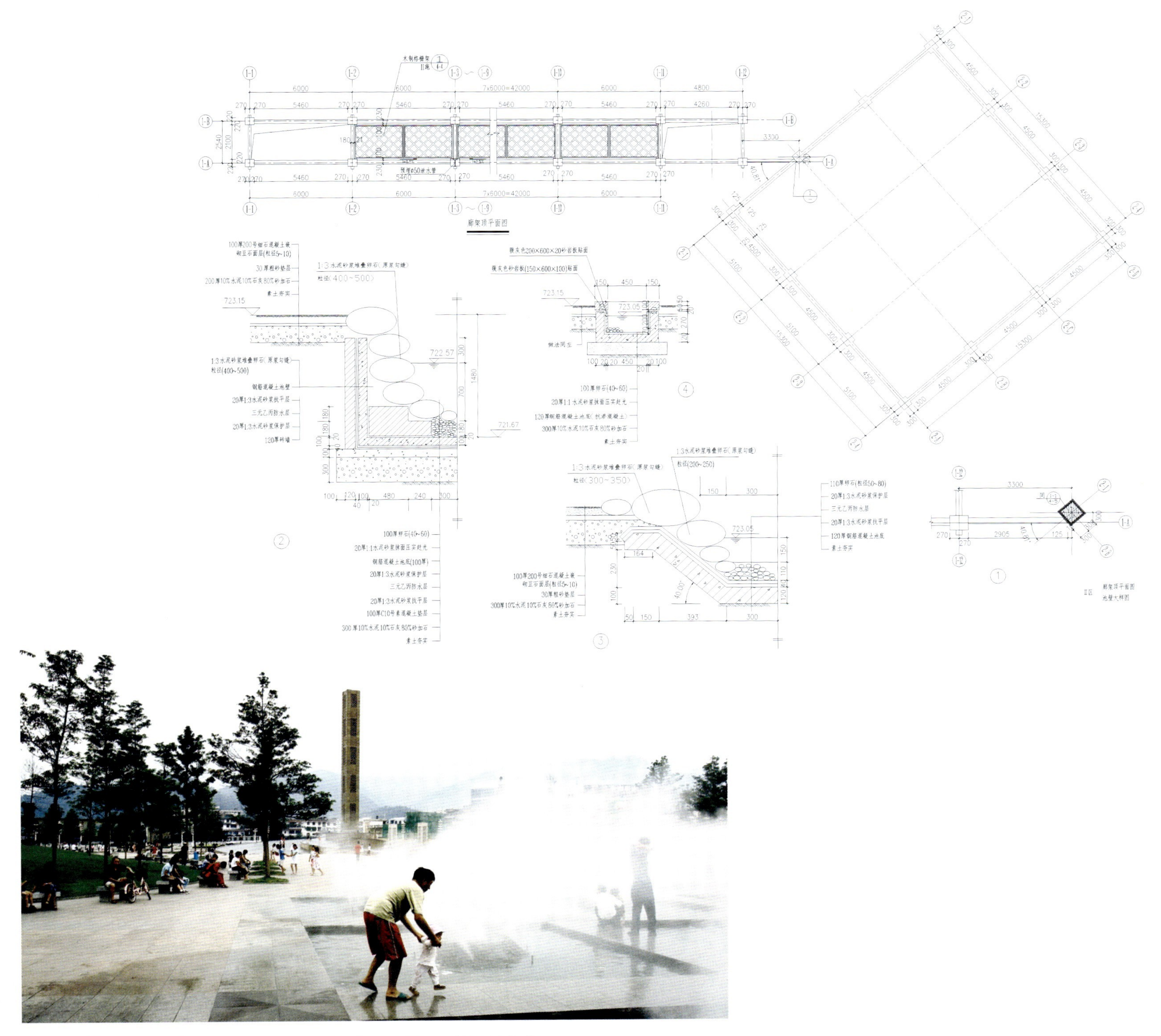

广场上的雾泉为夏日的城市带来凉爽，也营造了广场的喜剧性和神秘感
The mist fountain brings cool, dramatic and  misty atmosphere to the urban square

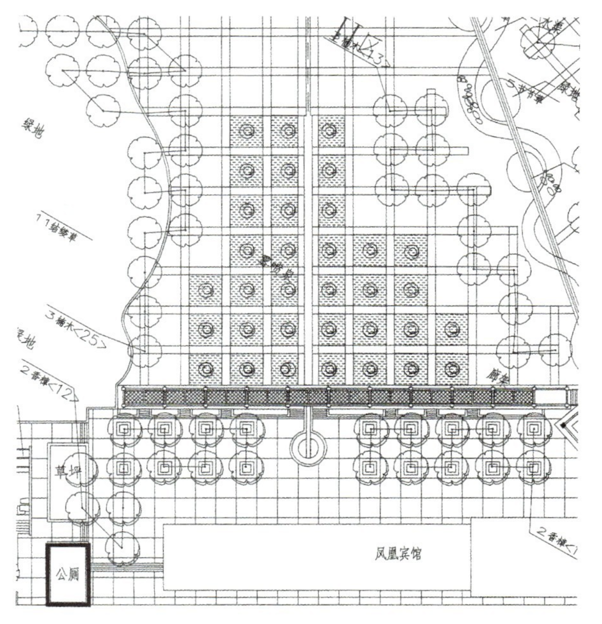
凤凰宾馆
公厕

临河休闲步道与大量条石座凳相结合，可以欣赏湍急的河流的涛声和白浪

The river walk with plenty stone seats, allowing people to listen the sound of weave and watch the white water

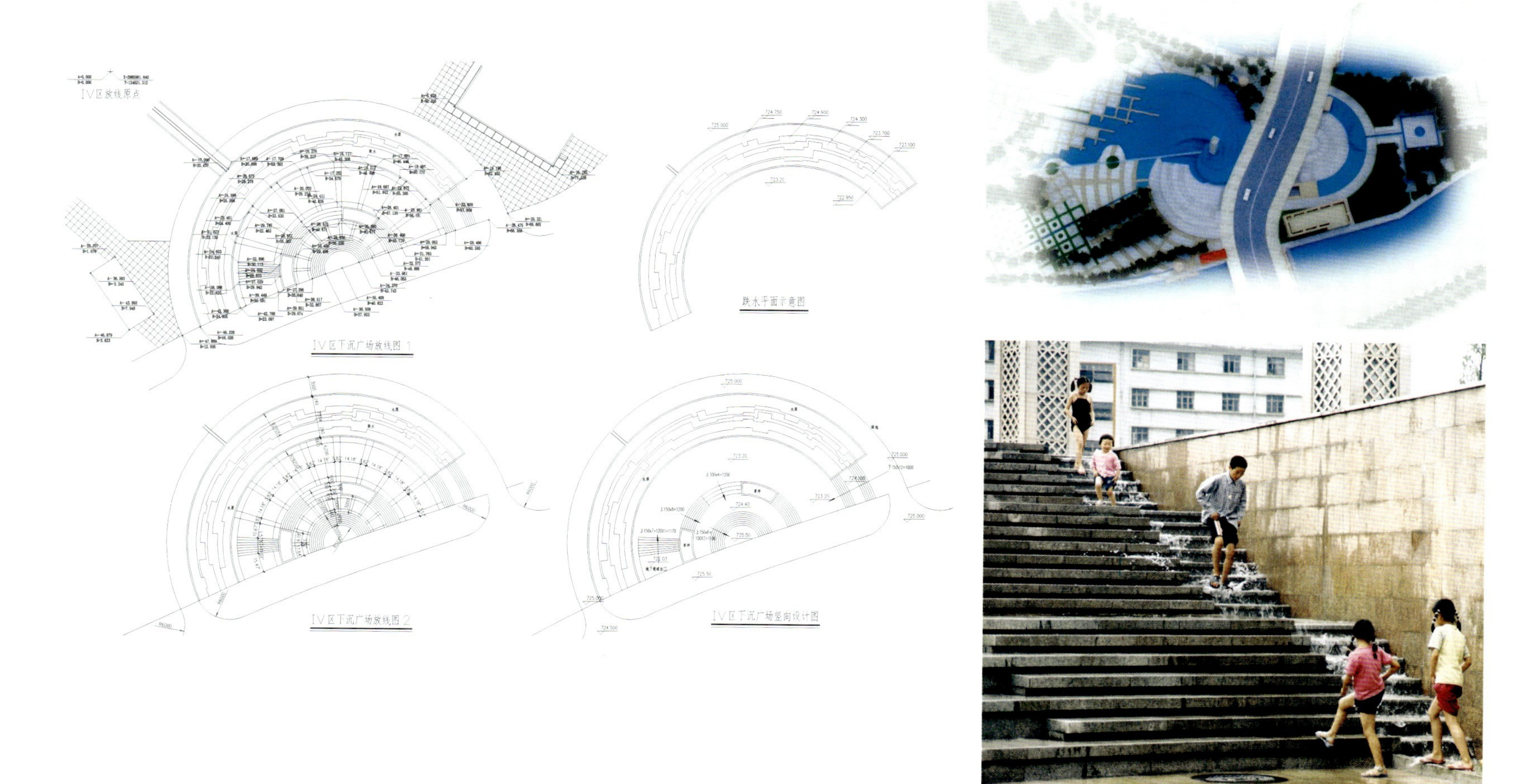

过街通道：联系II区和IV区，并将跌水景观结合在其中

The under pass between phase II and IV, cascade and fountain feature are integrated into the functional pass

桥：联系I、II、III区跨河桥是广场设计的有机组成部分，其形式语言同样从竹笼和当地的竹编工艺中获得启示

The bridge across the rivers between phase I,II and III. The formal language was inspired from the bamboo baskets

人性化设计：树池与座凳结合的设计提供大量林下休息场所；旱地喷泉的设计，使儿童们可以参与戏水
Design with people in mind: seats under the canopy, fountains on pavement provide opportunity as playground for people

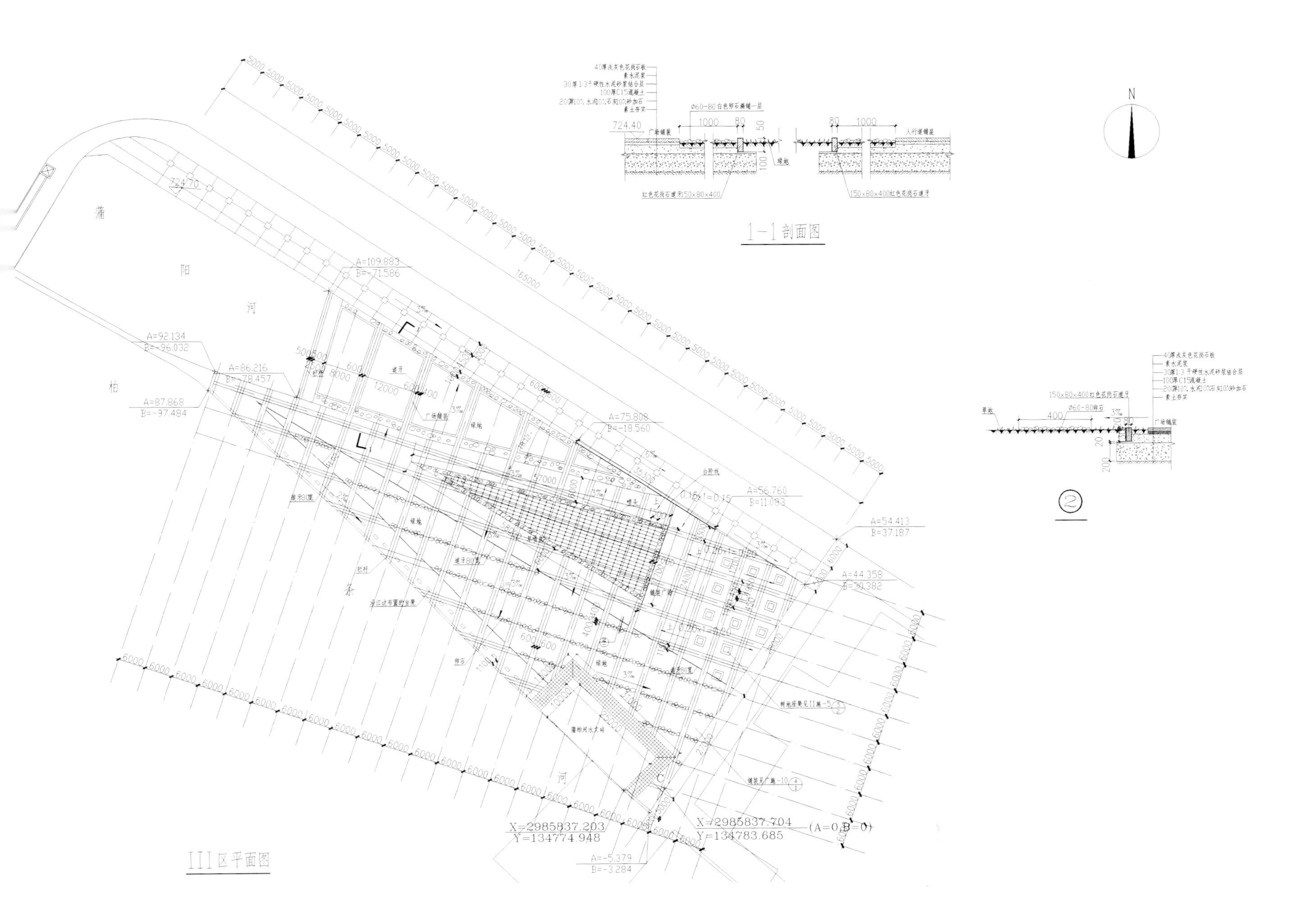

1—1 剖面图
N
②
III区平面图
蒲
阳
河
柏
茶
河
165000
A=109.883
B=-71.586
A=92.134
B=-96.032
A=86.216
B=-78.457
A=87.868
B=-97.484
A=75.808
B=-18.560
A=56.760
B=11.083
A=54.413
B=37.187
A=44.358
B=30.382
X=2985837.203
Y=134774.948
X=2985837.704
Y=134783.685
(A=0,B=0)
A=-5.379
B=-3.284
724.40
724.70

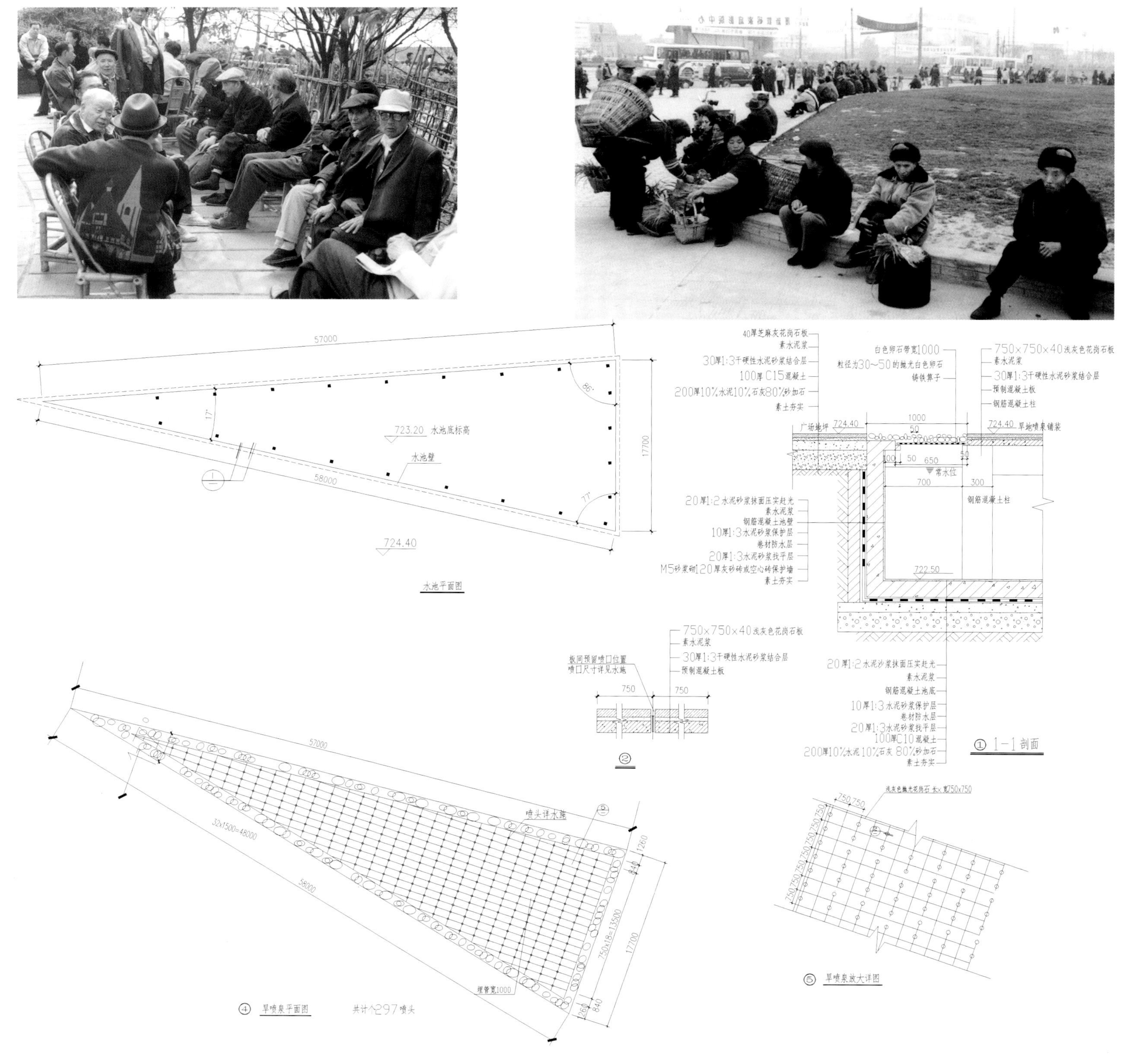

57000
58000
86°
77°
17°
17700
723.20 水池底标高
水池壁
724.40
水池平面图
40厚芝麻灰花岗石板
素水泥浆
30厚1:3干硬性水泥砂浆结合层
100厚C15混凝土
200厚10%水泥10%石灰80%砂加石
素土夯实
白色卵石带宽1000
粒径为30~50的抛光白色卵石
铸铁箅子
750×750×40浅灰色花岗石板
素水泥浆
30厚1:3干硬性水泥砂浆结合层
预制混凝土板
钢筋混凝土柱
广场地坪 724.40
724.40 旱地喷泉铺装
1000
50
100
650
常水位
700
300
钢筋混凝土柱
722.50
20厚1:2水泥砂浆抹面压实赶光
素水泥浆
钢筋混凝土池壁
10厚1:3水泥砂浆保护层
卷材防水层
20厚1:3水泥砂浆找平层
M5砂浆砌120厚灰砂砖或空心砖保护墙
素土夯实
20厚1:2水泥沙浆抹面压实赶光
素水泥浆
钢筋混凝土池底
10厚1:3水泥砂浆保护层
卷材防水层
20厚1:3水泥砂浆找平层
100厚C10混凝土
200厚10%水泥10%石灰80%砂加石
素土夯实
① 1-1剖面
750×750×40浅灰色花岗石板
素水泥浆
30厚1:3干硬性水泥砂浆结合层
预制混凝土板
板间预留喷口位置
喷口尺寸详见水施
750
②
喷头详水施
32×1500=48000
750×18=13500
1250
840
埋管宽1000
④ 旱喷泉平面图
共计个297喷头
浅灰色麻光花岗石 长×宽750×750
⑤ 旱喷泉放大详图

## 4.4 Ⅳ区图解 Phase-IV

盒子作为人的空间：在广场的北端，有一系列5m左右见方的、由石座凳围成的空间，它们是根据场地原有的一些揽船用的水泥坑的尺度设计的，它们曾经是、现在更是一家人、或是打牌和观牌人的适宜场所。生活和休闲方式本身给这些空间的设计带来了灵感

Boxes as people space: A view at the north end of the square, out door spaces were designed for people to site and enjoy family life as the local people usually do. Designs of these spaces were inspired by the experience of the living and leisure style of the local residents

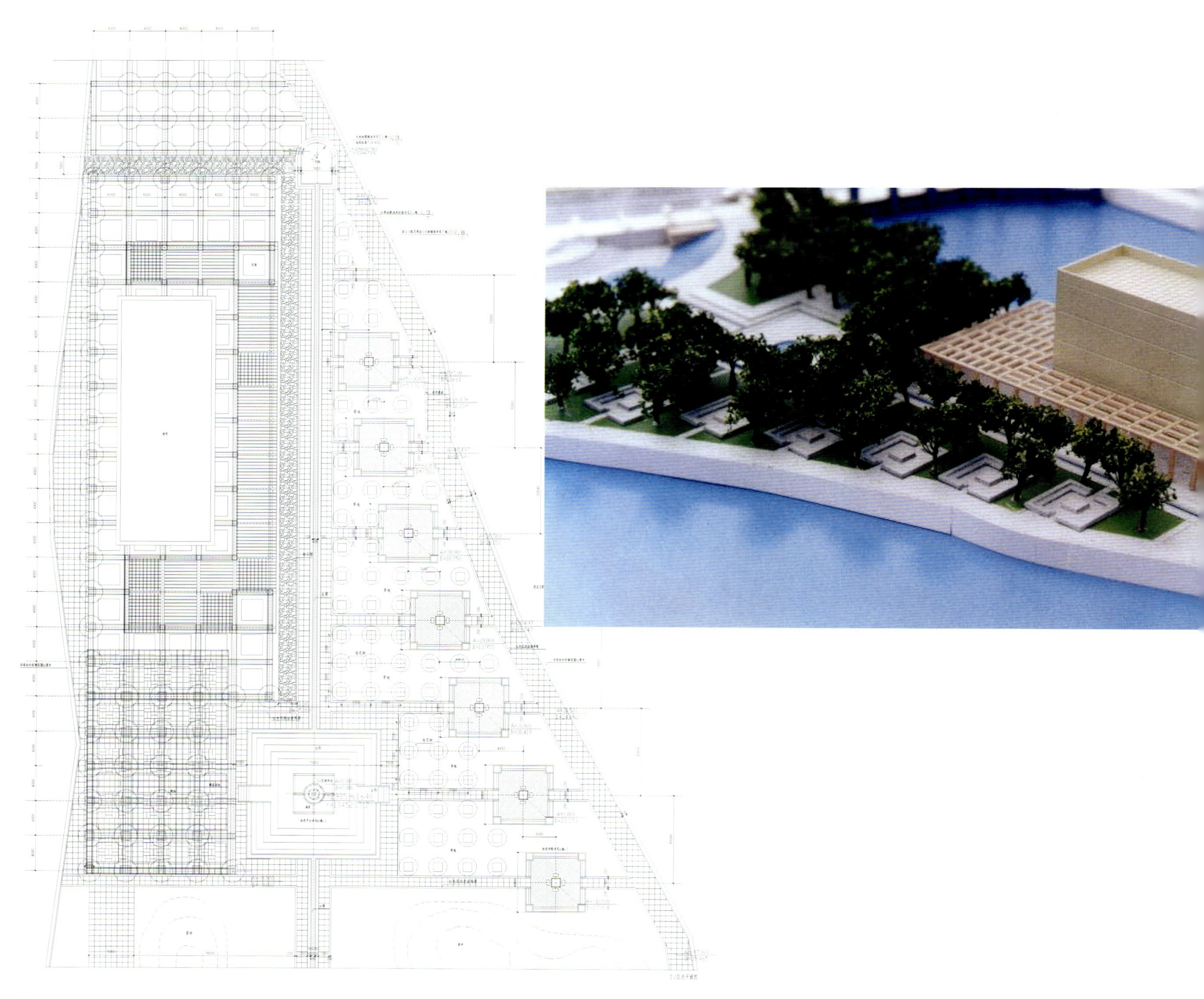

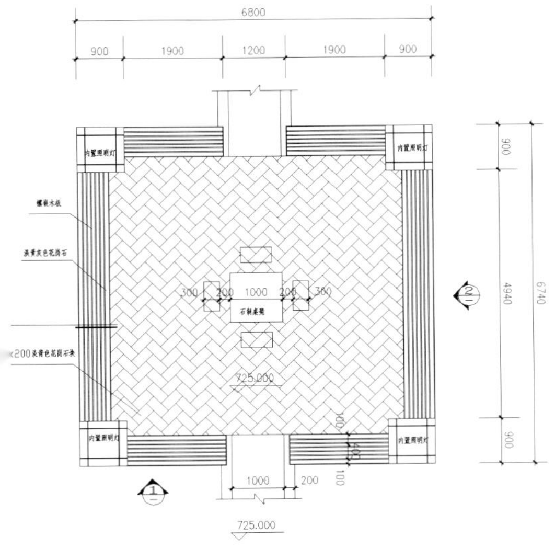
6800
900
1900
1200
1900
900
内置照明灯
725.000
石制座凳
1000
200
4940
6740
725.000
下沉井院平面图

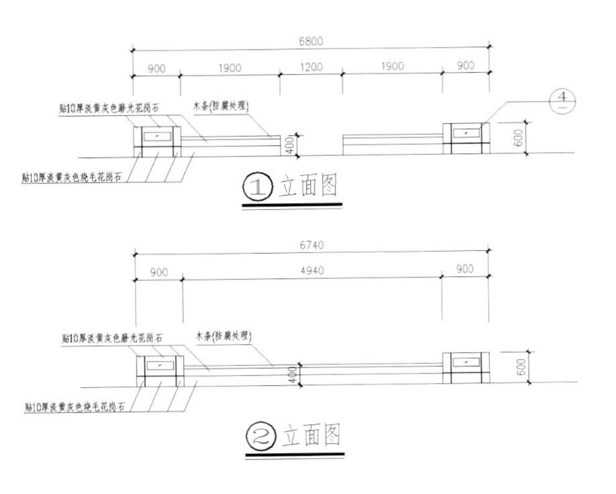
6800
900
1900
1200
1900
900
贴10厚淡黄灰色磨光花岗石
木条(防腐处理)
贴10厚淡黄灰色烧毛花岗石
400
600
①立面图
6740
900
4940
900
贴10厚淡黄灰色磨光花岗石
木条(防腐处理)
贴10厚淡黄灰色烧毛花岗石
②立面图

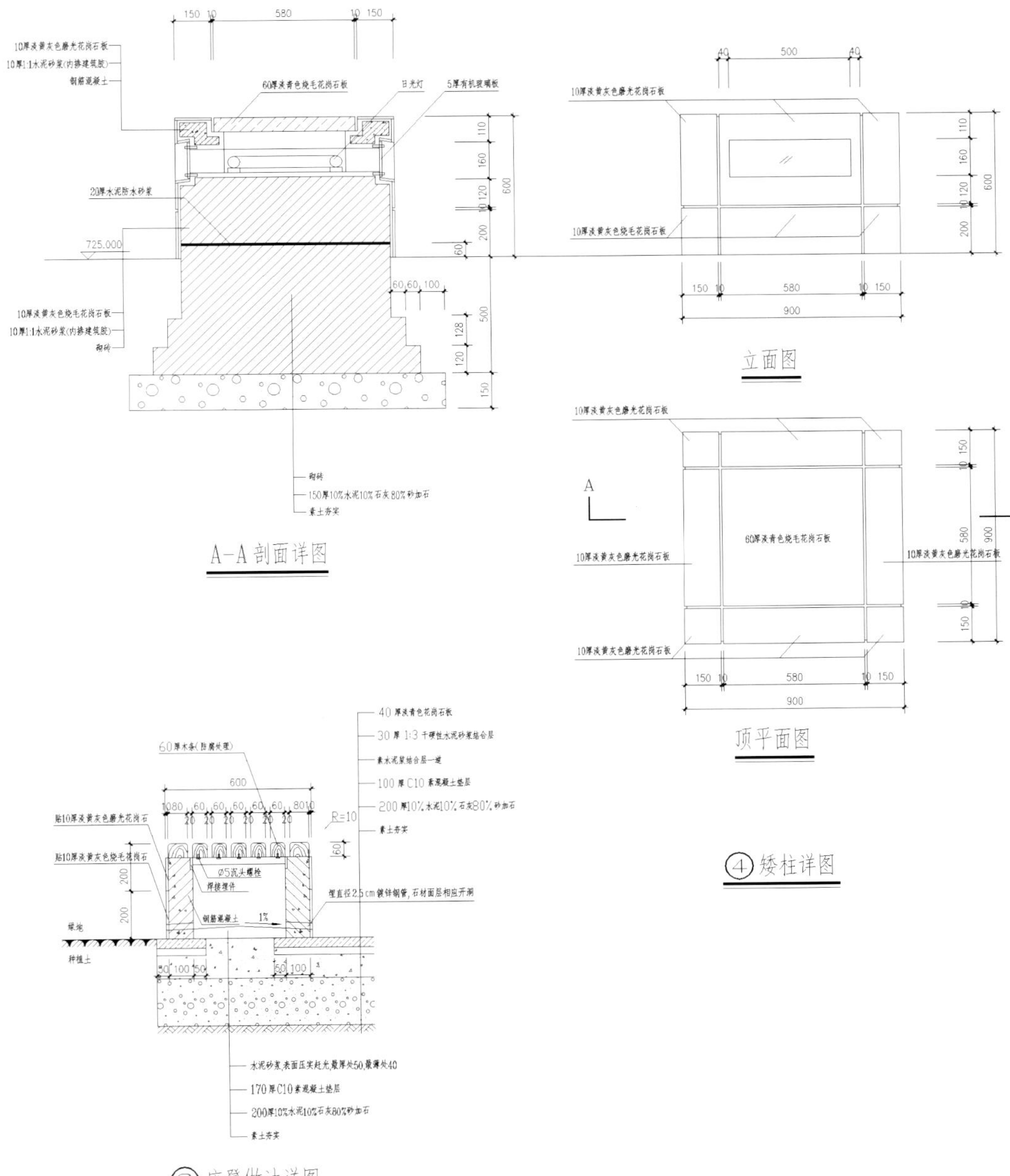
150
580
150
10厚淡黄灰色磨光花岗石板
10厚1:1水泥砂浆(内掺建筑胶)
钢筋混凝土
60厚淡青色烧毛花岗石板
日光灯
5厚有机玻璃板
20厚水泥防水砂浆
725.000
10厚淡黄灰色烧毛花岗石板
10厚1:1水泥砂浆(内掺建筑胶)
砌砖
150厚10%水泥10%石灰80%砂加石
素土夯实
A-A剖面详图
10厚淡黄灰色磨光花岗石板
10厚淡黄灰色烧毛花岗石板
500
900
立面图
60厚淡青色烧毛花岗石板
A
顶平面图
④矮柱详图
60厚木条(防腐处理)
600
40厚淡青色花岗石板
30厚1:3干硬性水泥砂浆结合层
素水泥浆结合层一道
100厚C10素混凝土垫层
200厚10%水泥10%石灰80%砂加石
素土夯实
贴10厚淡黄灰色磨光花岗石
贴10厚淡黄灰色烧毛花岗石
R=10
Ø5沉头螺栓
钢筋混凝土
1%
种植土
水泥砂浆,表面压实赶光,最厚处50,最薄处40
170厚C10素混凝土垫层
200厚10%水泥10%石灰80%砂加石
素土夯实
③座凳做法详图

## 土人景观设计著作系列

为推动景观设计学科的理论和实践在中国的发展，北京大学景观设计学研究院、北京土人景观规划设计研究所与中国建筑工业出版社等合作，连续出版理论专著、实践案例和译著，近年来已出版的著作包括：

俞孔坚，王建，黄国平，土呷，李伟，陀罗的世界：东乡土景观阅读与城市设计案例．北京：中国建筑工业出版社，2004

俞孔坚，刘向军，李鸿．田——人民景观叙事南北案例．北京：中国建筑工业出版社，2004

俞孔坚，庞伟．足下文化与野草之美——岐江公园案例．北京：中国建筑工业出版社，2003

俞孔坚，李迪华．景观设计：专业，学科与教育．北京：中国建筑工业出版社，2003

俞孔坚，李迪华著．城市景观之路．北京：中国建筑工业出版社，2003

俞孔坚，Davorin Gazvoda，李迪华等著．多解规划——北京大环案例．北京：中国建筑工业出版社，2003

俞孔坚等著．高科技园区景观设计——从硅谷到中关村．北京：中国建筑工业出版社，2001

俞孔坚著．景观：文化、生态与感知．北京：科学出版社，1998，2000

俞孔坚著．景观：文化、生态与感知．中国台北：田园文化出版社，1998

俞孔坚著．理想景观探源：风水与理想景观的文化意义．北京：商务印书馆，1998，2000

俞孔坚著．生物与文化基于上的图式——风水与理想景观的深层意义．中国台北：田园文化出版社，1998

俞孔坚编著．设计时代．石家庄：河北美术出版社，2002

周年兴，李小凌，俞孔坚等译．(F．Steiner 原著)．生命的景观——景观规划的生态学途径(第二版)．北京：中国建筑工业出版社

孟亚凡，俞孔坚等译．(C.Birnnaum 和 R.Karson 原著)．美国景观设计先驱．北京：中国建筑工业出版社，2003

俞孔坚，王志芳，孙鹏等译．(J.Simonds 原著)．景观设计学——场地规划与设计手册．北京：中国建筑工业出版社，2000

俞孔坚，王志芳，孙鹏等译．(C.Marcus 和 C.Francis 原著)．人性场所．北京：中国建筑工业出版社，2001

刘玉杰，吉庆萍，俞孔坚等译．(N.Nines 和 K.Brown 原著)．景观设计师简易手册．北京：中国建筑工业出版社，2002